DR SPACE JUNK
VS
THE UNIVERSE

ALICE GORMAN is an internationally recognised leader in the field of space archaeology. She investigates the archaeology and heritage of space junk, planetary landing sites, rocket launches and antennas. She is a Senior Lecturer at Flinders University in Adelaide, and a Director on the Board of the Space Industry Association of Australia. In 2017 she won the Bragg UNSW Press Prize for Science Writing. Dr Gorman tweets as @drspacejunk and blogs at Space Age Archaeology.

*This book is dedicated to the women of my family
across four generations:
my grandmother, Alice, my mother, Tish,
my sister, Claire, and my niece, Stella.*

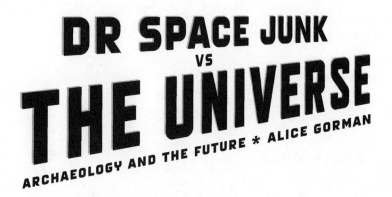

DR SPACE JUNK
vs
THE UNIVERSE

ARCHAEOLOGY AND THE FUTURE * ALICE GORMAN

The MIT Press
Cambridge, Massachusetts
London, England

An earlier version of the work was first published in Australia by NewSouth, an imprint of UNSW Press Ltd.

This book was set in Granjon by Josephine Pajor-Markus. Printed and bound in the United States of America.

Library of Congress Cataloging-in-Publication Data

Names: Gorman, Alice (Alice Claire), author.
Title: Dr Space Junk vs the universe : archaeology and the future / Alice
    Gorman.
Other titles: Doctor Space Junk versus the universe
Description: Cambridge, MA : The MIT Press, [2019] | Includes bibliographical
    references and index.
Identifiers: LCCN 2019017900 | ISBN 9780262043434 (hardcover : alk. paper)
Subjects: LCSH: Space archaeology. | Space debris. | Cultural property.
Classification: LCC TL788.6 .G67 2019 | DDC 629.409/009—dc23 LC record
available at https://lccn.loc.gov/2019017900

10   9   8   7   6   5   4   3   2   1

CONTENTS

FOREWORD

There's something nicely counterintuitive about the concept of space archaeology. We generally think of archaeologists as people who dig *down*, trowelling through the topsoil and uncovering the deeper past beneath. Alice Gorman and her ilk dig, as it were, *up* – into the heavens above the breathable sky – as well as into the dirt, to extract the debris that has fallen, Lucifer-like, from the heavens. History is the past, and archaeology is often assumed to be the deep past, but many people still think of spaceflight as science fiction and so therefore, to some degree, futuristic. And perhaps most strikingly of all we think of archaeology as *slow*: the painstaking scraping away of layers, brushing dust from a half-cracked bowl. Space is, whatever else it is, *fast* – we need to be moving at 40 000 kilometres an hour just to get there, and a body in orbit will be zooming along at somewhere between 24 000 and 32 000 kilometres an hour, depending on the kind of orbit we're talking about.

In Alice Gorman's wonderful book, the world is turned

upside down. The past is above us rather than below. The immensely expensive and cutting-edge technologies of Apollo are junk. She takes us on a marvellous odyssey through the Space Age in terms of what it has left behind, rooting her expertise in her own life experience as well as in the grander narratives of the Cold War struggles and titanic engineering feats.

As Dr Gorman notes it's wrong to think that the term 'history' only applies to temporal remoteness. 'Heritage,' she says, 'is about things from the past which are significant to people in the present, and which they want to keep into the future. This applies to very recent things like space exploration as much as it does to ancient rock art.' History means everything that's been that has any kind of resonance to the present. It situates, speaks to and explains the nature of the present. There are many reasons why historical amnesia is an evil, but a main one is that such forgetfulness allows the unscrupulous to replace history with mythology. Myths make good stories, but because they are not true they are easily bent to malignant ideological purposes. There is an absolute need to keep the past alive in the present, both to avoid its mistakes and – particularly where the glories of our human Space Age are concerned – to remind us of what we were once capable. If ever it feels like contemporary humanity is stumbling, or perhaps shuffling backwards, it is salutary to remember that within living memory we as a species were capable of striding *all the way to the Moon*. It makes the hairs prickle on the back of my neck just to think about it. If I am certain of one thing about that troubled century, the twentieth, it is that in 10 000 years, when every other name has been forgotten, every politician and celebrity and artist consigned to the footnotes of specialist historical accounts, Neil Armstrong's name will still be common currency.

One thing space archaeology tells us is that the past is quicker, and more immediate, than we think it is. The Space Age sits at the moment in an uncanny valley of memory: too recent to be thought of as 'history' in the Ancient Greek or Pharaonic Egypt sense of the word, but too far back in time to inspire the young generation. For a long time, going to the Moon was the stuff of the imagined futures of science fiction. Then, in 1969, all sci-fi's dreams came suddenly true. And then again, blink or you might miss it, the dreams hurtled backwards, at 32 000 kilomtres an hour, into our collective past. Going to the Moon is no longer what we do; it's what our ancestors used to do. An age science fiction predicted would last for thousands of years of human space exploration and colonisation was, in reality, compressed into a few years, packaged away and shut down. That historical moment, once blue-shifted by its approach down the speculative timelines, is now red-shifted by its rapid retreat from us into posterity. It grows more distant every day.

Gorman neither over-romanticises nor dismisses the junk she champions. It can't all be conserved – but most space junk is constantly moving, on a spiral that will bring it careering into Earth's atmosphere at speeds that are, literally, meltingly fast. She sanely cites the Burra Charter, to do 'as much as is necessary and as little as possible' to retain cultural significance. 'Every family farm has its machine graveyard and rubbish tip. It's OK to let them decay, to let the wind and the rain take them.' Some space junk may pose dangers for future astronauts (as was so vividly portrayed in Alfonso Cuarón's Oscar-winning 2013 movie *Gravity*); but 'we don't need to destroy everything currently classed as space junk either, over 95 per cent of all the stuff up there, to reduce the risks of collisions from orbital debris. We can do it in a smart way by thinking through all the heritage and environmental issues.'

This book reminds us that outer space is not impossibly distant: that it's almost close enough to touch, and that it often showers objects down upon us – like the fragment of Skylab she keeps on her desk. As charming as it is expert, as gripping as it is surprising, *Dr Space Junk vs The Universe* deftly threads together the cosmic and the personal, the stupendousness of space with the lived experience of human beings down here, from Aboriginal Australians to 21st-century city dwellers, encompassing everything from astounding space technology to cake and cocktail recipes. Outer space may be an arena of velocity, but this is a book you'll want to take your time with – to savour, rather than to gulp. It's a book that juxtaposes the human and the universal, and that's important, because it helps keep vital the great human dream, in the longest terms of our future as a species, the greatest of all of them: that one day we'll retake that step to the Moon, only this time we'll keep on going.

Adam Roberts

ACKNOWLEDGMENTS

This book has been an intense journey, and the support and encouragement of friends, family and colleagues has been so important. Chief among these is my mother, who instilled a love of words and ideas in me from my earliest consciousness.

My editors at *The Conversation*, Michael Lund, Sarah Keenihan, Belinda Smith, Paul Dalgarno and Matt de Neef, have been very influential in teaching me how to write for general audiences. Many sections of this book started out as *Conversation* articles where they allowed me free rein with ideas. Another significant influence has been participating in the Best Australian Science Writing community since 2014. In 2017, with the encouragement of Patrick Allington, I wrote an essay for the *Griffith Review* which was selected for the *Best Australian Science Writing* that year, and went on to win the Bragg UNSW Press Prize for Science Writing. This was a real turning point for me, and I'd like to thank Kathy Bail from UNSW Press, and Julianne Schultz from the *Griffith Review*, for their encouragement. Julianne introduced me to my agent Jane Novak. Working with Jane has been an education and I'm so glad to have had the benefit of her sage advice and literary acumen.

Phillipa McGuinness from NewSouth Publishing nudged this book from a collection of essays into a narrative. It's really nice having someone who believes in you! I appreciated her thoughtful comments and pep talks all the more as they coincided with the publication of her own book. Jocelyn Hungerford did an amazing job with copyediting, and Paul O'Beirne was always at the end of the phone if I needed him.

Patrick Allington has been a mentor and advisor throughout and I can't thank him enough for all the myriad ways he has helped me. Adam Roberts, one of my all-time favourite authors, was beyond generous with his advice and support (I bow down before him). My esteemed colleague Lynley Wallis was the purveyor of gins and tonic, cakes when required, and endless indulgence for my space nerdery.

Claire Gorman and Daniel Jordon offered invaluable advice at crisis points. Jaydeyn Thomas and Elizabeth Weeks gave me insightful feedback on earlier versions of this book and made me feel like I could really do it. For later versions, Ian Moffat and Anika Johnstone's enthusiastic and critical responses helped me keep on track. My space colleague Graziella Caprarelli from Hypatia Scientifica generously gave up her time to fact-check the science.

Many friends were willing to have their brains picked, answer stupid questions, or generally contribute to a positive environment for the writing to take place. For this I thank Jane Lydon, Heather Burke, Claire Smith, Tully Barnett, Heather Robinson, Karen Ashford, Katherine

Sutcliffe, Ceridwen Dovey, Ralo Mayer, Brett Biddington, Peter Nikoloff and Matthew Spriggs.

Others who played an important role include Paul Filmer of the US National Science Foundation, who runs a Twitter account for the Voyager 2 spacecraft (@NASA-Voyager2). Parts of chapter 6 are based on a beautiful interview he did with me, playing the role of the spacecraft, in 2012. *Anthropology News* enabled me to write about Rosetta and Philae. Eric Bouvet inspired me to think about engineering rings of debris around Earth. My work at the Orroral Valley Tracking Station was funded by the ACT Heritage Unit. The futurolinguistics exercise with gravity was developed with students at the University of Applied Arts in Vienna, and I thank Eleni Boutsika-Palles, Lara Erel and Rina Lipkind for their imaginative ideas. Space dancer Allegra Searle-LeBel introduced me to the concept of the Small Dance, which made a huge impression on me (as did she). Doug Villain helped me with calculations to convert space junk into the equivalent amount in cane toads. The piece of Skylab sitting on my desk was the kind gift of Kevin Hamdorf.

Thank you everyone, and I hope you enjoy the book.

INTRODUCTION:
LOOKING UP, LOOKING DOWN

I'll make my report as if I told a story, for I was taught as a child on my homeworld that Truth is a matter of the imagination.

Ursula K Le Guin, The Left Hand of Darkness

I was sitting on a bus in the early evening, on my way home from work. It was a clear bright warm day and I was looking out of the window at the sky, framed by occasional flashes of green from magnificent old trees in the rather swanky Adelaide suburbs I was passing through.

I was contemplating the blueness of the sky and thinking what a fortunate colour it was. An orange sky, as we might experience on Venus, would feel so much harsher to our moist terrestrial sensibilities.

The blue sky is wonderful, but in the day its curved dome prevents us from seeing outwards to the stars, as we can at night. During the day, when we're awake, it's like the lid is on our jar: we're looking down, inwards, at our feet and not above our heads. When the lid is raised at night, we're often inside, and then asleep until dawn. Even if we are outside, light pollution from cities, towns and industry has obscured so much of what we can see in the night sky. So we're not always conscious of the vasty deeps of the solar system and interstellar space. We don't feel ourselves part of space.

So different from fish, I thought to myself. In the water, you'd always be conscious of *up*, where the pressure was lower, and of the direction of the light as it filtered down towards the sea floor. I imagined myself on the bottom of a shallow sea, looking up at the sky as if it were the surface of the water, the clouds as the underside of white foamy waves.

Thinking myself into this perspective made me feel sort of opened out, expanded. I could feel the presence of space so much more. It was actually a physical sense, as if someone had lifted a heavy weight from my head – I was still moored to the surface, but I could feel the growing lightness above me. It was different from the marvellous lack of gravity that accompanies flying dreams. Earth was definitely still in control here. I just had an awareness of being at the bottom of something much larger, instead of being on the surface of something much denser. What would it be like if this were our natural way of feeling? Our feet rooted in the soil, but our mind's eye always attuned to the heavens, an antenna of flesh and bone?

I have to make an effort to feel this openness to the

cosmos, but for more than a decade now I've lived between these two worlds as an archaeologist who investigates space exploration. From walking always with my eyes focused on the ground, reading the traces of the past in the stone tools discarded by people long before I was born, I've taken to looking up towards the less visible material traces of a present that bounces back and forth between Earth and space. I'm concerned with the material and social dimensions of space exploration, rather than just the technology or the science. I'm not just interested in the engines that enable a rocket to reach space: I'm also curious about the place the rocket was launched from, the people who worked on it, where it ended up, and how it's been remembered. Dr Space Junk is my alter ego, a nickname given to me years ago which has stuck. She is part of my professional persona now; I'm known by this name in online spaces and often in real life as well. Dr Space Junk is the bridge between these two things that, on the surface, don't seem to go together – space exploration and archaeology.

ON EARTH AS IT IS IN HEAVEN

In July 1969, the whole world was looking up. We watched three Apollo 11 astronauts travel to the Moon in a cramped capsule, in the hope they'd become the first people to set foot on another celestial body. It was so risky that US president Richard Nixon had a speech already prepared in case the astronauts did not survive.

At the same time as this mission, German archaeologist Wolfgang Erich Wendt was excavating in a cave in south-western Namibia, in the south of the African continent. Wendt and his team were captivated listening to the radio coverage of Apollo 11. As they dug down through the layers of the past, the future was being made over 300 000 kilometres, and another world, away. When he heard that the astronauts had returned safely to Earth, Wendt named the cave Apollo 11 in their honour. The new archaeological site of Tranquility Base on the Moon, created by the artefacts the Apollo 11 astronauts left behind, was mirrored by an ancient site on Earth.

And not just any site. One of the aims of Wendt's excavation program was to find a chronological sequence of cultural changes in the region, so it was important to get accurate dates. The cave was painted with red and white bees and geometric patterns, but rock art on walls is notoriously difficult to date. However, the team got lucky at this site. Seven stone plaques – flat slabs of grainy rock – with white, charcoal and ochre paintings of animals on them were excavated from the accumulated dirt that filled the cave, and so they could be dated by association. Charcoal in the same layer returned radiocarbon dates of 27 500–25 500 years before present (bp). More recent dates, using a method called Optically Stimulated Luminescence, show that the plaques are about 30 000 years old.

The animals depicted included an elegantly elongated zebra, a black rhinoceros, and a figure which looks like a cross between a big cat and an antelope. Because of its slightly odd hind legs, some say it is also a 'therianthrope' – a human/animal hybrid known in rock art from other parts of the world, and associated with rituals and trance states. Earlier geometric and abstract art has since been found, but these plaques are still the oldest figurative art in Africa.

The date matters because rock art is not just about the artistic life of ancient people; it's an indication of symbolic behaviour – a prerequisite for language, art and spiritual beliefs. It's part of a suite of characteristics known as behavioural modernity, which is argued to be the foundation of contemporary human society and technology. The suite includes making cultural innovations, complex technologies, advanced problem solving, and long-range planning abilities. It's pretty clear that behavioural modernity existed by the time people were occupying the Apollo 11 Cave, at 30 000 years ago. The question is, how long before this did it first emerge? To answer that, we need to find the earliest archaeological evidence of using symbols to communicate, as well as evidence of thinking about the future.

What would this archaeological evidence look like? Skeletal remains, particularly those showing changes in brain structure imprinted on the inside of the skull, and bones in the palate and throat which suggest the range of sounds that could have been made, tell us that language is physically possible, but not that it occurred. If you change or modify your body to communicate something to others, as in tattoo or scarification, it's very symbolic – but soft tissue seldom survives the ravages of time. Beads and personal ornaments made out of stone, bone, shell and teeth may survive in the archaeological record, however, and they also show an intent to communicate, whether that is status, identity or mood. Standardised artefacts distributed over large areas, rather than ad hoc ones made randomly in the moment, indicate that people communicated through the style of manufactured objects. These artefacts might be a type of stone tool such as the pointed, flaked Acheulean hand axe, or something like the Ice Age Venus figurines, found across Europe. Technology that requires many steps over a period of time indicates planning for the future as well as the ability to remember the past. Such technology might involve securing the raw materials to make an artefact over days, weeks or months: the right piece of wood for a handle; tree resin, ochre and animal fat to make an adhesive; stone to make a sharp flake just the right size and shape; a fire to melt and mix the resin ingredients. Finally, the stone flake is hafted on its handle and ready to use for when you need it – a future certainty which has not yet occurred.

The art in the Apollo 11 Cave in Namibia is symbolic. The painted animals are not only symbols of the living ones, but represent a belief system where painting them had some impact in the world. They also relate to another cultural process with very modern resonances. Cognitive scientist Andy Clark calls humans 'natural born cyborgs', arguing that human life has always combined biological and technological components. The plaques could be interpreted as a form of memory storage outside the brain. This is memory made of stone, a touchstone perhaps, to recall – well, we don't know what. The software inside the brain is gone and there's only the hardware left in the form of artefacts and bones.

It used to be thought that only *Homo sapiens* displayed behavioural modernity, but now the picture is not so clear. Archaeological research reveals more and more about the complexity of Neanderthal life, and new hominin species like 'the Hobbit' in Indonesia and the Denisovans in Siberia have disrupted the old, established human lineages. It's no longer a straight evolutionary path from primate ancestor to modern human, as you see it represented in textbooks and on T-shirts, where the knuckle-walking ape gradually bends upright, becomes bipedal and morphs into a striding human (usually male). Sometimes, an astronaut is placed at the end of the sequence. The astronaut is perhaps the ultimate cyborg, a body interfaced with technology whether inside a suit or a space station, and constantly monitored to stay in the state of equilibrium which keeps it alive.

What Wendt's team had found was, at the time, momentous. Here was the earliest evidence that *Homo sapiens* in Namibia were thinking symbolically. As Wendt listened to the broadcast of Neil Armstrong setting his foot down on the dusty lunar surface, was he also thinking about human evolution and the stage that space exploration represented?

Perhaps he pondered how the symbols of the Apollo lunar landing site compared to the stone plaques. The landing module, which was left behind on the lunar surface, was named the 'Eagle', a national emblem of the US. The command module, which remained in orbit around the Moon with one astronaut on board during the mission, was called 'Columbia' after the cannon Jules Verne used to launch his fictional Moon rocket in 1865. The name also alluded to the voyage of Christopher Columbus to the Americas in 1492, considered a foundational act in the establishment of the US as a nation. The most obvious symbolic artefact was the US flag planted in lunar soil, a well-known metaphor (and cliché) of colonisation.

The terrestrial Apollo 11 site was located in the traditional country of the Nama people. They were the victims of what has been called the first genocide of the 20th century, when they rebelled against German colonial rule in Namibia. Between 1904 and 1908, 80 per cent of the Nama and Herero people were killed and the rest put in concentration camps. It's not known what those who survived thought of the lunar landing, but at the same time as 'one giant leap' for humanity, a nasty version of the eye infection conjunctivitis was doing the rounds in Ghana, Sierra Leone, Nigeria and other West African nations. People called it Apollo – a word that rolled more easily off the tongue, some opined, than the English term conjunctivitis. The infection was frequently attributed to the arrival on Earth of lunar dust stirred up by the astronauts, who had themselves been quarantined on return so they would not introduce unknown lunar pathogens to Earth.

The astronauts did bring dust back to Earth, but as scientific samples. It became a valuable commodity, even more precious with every passing year that humans did not return to the Moon. For those proposing to mine the Moon for its water, rare earth elements and helium-3 fifty years later, this dust is more than symbolic evidence of lunar conquest.

It holds secrets about geology and soil mechanics: the new challenges of a lunar exploration looking to exploit the resources of the Moon.

A NEW ERA OF SPACE

Human ancestors left their cultural footprint in the red sands of Namibia 30 000 years ago, and now that footprint has been translated into interplanetary space, from the teeming satellites in Earth orbit, and landing sites on the Moon, Mars and Venus, to the Voyager spacecraft at the edge of the solar system. As we nudge towards the third decade of the 21st century, access to space is slowly moving out of the hands of national governments with the rise of commercial spaceflight development, and the growth of the space tourism market.

Small, off-the-shelf satellites are making access to Earth orbit more and more affordable, as the costs of building and launching a space instrument come down. At the same time, there's a new breed of space messiah. Astronauts are no longer the only heroes of space. Instead, cults are forming around extremely wealthy entrepreneurs, like Elon Musk, Jeff Bezos and Richard Branson, who are developing commercial space transport systems for tourists and colonists. It's shaping up to be an era of grand gestures and grand investments from the private sector – as we saw with the launch, in February 2018, of Elon Musk's red sports car into solar orbit. It seems that more people are thinking about space than we've seen since the Apollo era of lunar exploration ended in the early 1970s. Suddenly space is sexy again.

Well, perhaps it's not so sudden. What is happening now has been building up for more than a decade. There have been epic deep space missions like the European Space Agency's Rosetta spacecraft, launched in 2004 to make the first ever landing on a comet in 2015. Then there was NASA's New Horizons, launched in 2006, which showed us the surface of dwarf planet Pluto for the first time in 2015 and is now moving out into the Kuiper Belt, the far home of icy comets. Closer to our home in the inner solar system, people are not just speculating about creating human settlements on Mars and the Moon, they're planning them in all earnestness. Space tourism, once only possible for the obscenely wealthy, is starting to look within reach for the moderately wealthy, and perhaps soon for the regular person. As former astronaut trainer Dr Walter Peeters has said, 86 per cent of Earth's population is physically fit to travel to space – it's just that it's so expensive, and there are no facilities for people not trained to be crew. On the commercial side, companies have been set up based on the potential market for asteroid and lunar minerals, while space scientists research the considerable practical challenges of extracting minerals from other celestial bodies.

As a result, the United Nations space treaties, which were established during the Cold War to prevent territorial claims being made over space, are now being questioned and debated. Many wonder if the existing framework is capable of dealing effectively with technologies and industries which were only dreamt of in 1967 when the Outer Space Treaty (OST) was opened for signature. One of the key principles of the OST is that space is the common heritage of humanity and cannot be owned by anyone – government, nation, individual or corporation. Space is very colonial: we talk of the 'conquest' of space, the 'high frontier' or the 'final frontier', colonising other planets, and the innate urge of human beings to explore, often without thinking about it; it's such a strong master narrative. Instead of considering the treaty to be outdated, we might equally think of it as a radical statement of equality and justice – and one we need more than ever.

This is a new version of an old story precipitated by the Cold War 'Space Race', when the US and the USSR competed to be first in space. It's really about who gets to imprint space with meaning. Sure, it's also a story about technology, science and industry, and taking humanity to the stars – but at the end of the day, what's really at stake is who gets to control the narrative of space. The Cold War narrative, put very crudely, was about capitalism vs communism. Space is no less an ideological battleground today, but now, it's about extreme late capitalism vs digital democracy. Or private good vs public good, or military vs civilian. We have a chance to make a new world beyond this world, but it will be the same old world all over again if we're not paying attention to the narrative. It won't be different if it's supporting the same terrestrial power structures as before. It's really about how we connect the past with the future in space. And for that, understanding heritage is essential.

DR SPACE JUNK'S TOUR OF THE SOLAR SYSTEM

Heritage is about things from the past which are significant to people in the present, and which they want to keep into the future. This applies to very recent things like space exploration as much as it does to ancient rock art. I used to be a heritage consultant, working with Aboriginal communities in Australia. It's natural for me to ask different kinds of questions about the history and impacts of space than those that might be obvious to an aerospace engineer, for example. Space narratives usually leave out Indigenous people and often 'non-spacefaring' nations too – which is a large chunk of the world. We can't afford to do that any longer, not if we're truly committed to space being for *all* humanity. I'm an insider in the space world – I no longer get strange looks from people wondering what the hell an archaeologist is doing here! I'm not looking in from the outside and being critical – this is my community too, and I want it to be the best it can be.

We could look at it another way, from our vantage point of this decade in the 21st century. From the origins of the V2 rocket in the Second World War until the end of the Cold War in 1991, space exploration was driven by conflict and national prestige. We've had sixty years of space culture rooted in war, and only thirty years of space culture driven, on the surface at least, by peace. Despite having had a UN treaty on the peaceful uses of outer space since 1967, you could say we haven't yet learnt how to be peaceful in outer space. This means the archaeology of space is no optional luxury, but a necessity. The ability to interact with space has had a profound impact on how we live on the surface of Earth, and if humans are going to make good decisions about the future, then we need to understand everything we can about how the objects and technology of space structure the everyday world. For an archaeologist, this technology is a window into the world view that produced it, not just while it was in use, but after it was discarded or abandoned as well.

In this book, I want to take a physical journey through the solar system and beyond, and a conceptual journey into human interactions with space. My tools are artefacts, historical explorations, the occasional cocktail recipe, and the archaeologist's eye applied not only to the past, but the present and future as well. This is about understanding the ways we make space meaningful.

Sitting on my desk is an artefact that always makes me smile. It doesn't look like much. It's a narrow sliver of blond bamboo-textured material about 10 centimetres long and 1 centimetre wide. Carefully handwritten on it in black ink are the words 'A piece of the NASA Skylab spacecraft crash landed in Western Australia July 11th. Collected in San Francisco California 1979.' The ink has bled into the fibres, blurring the letters, but they're still mostly legible.

Skylab was the first US space station, launched into Low Earth Orbit in 1973. Astronauts lived in it, off and on, for 168 days. They ate frozen Lobster Newburg and made a Christmas tree out of empty food cans. They went on strike when the work demands of mission control became too much for them. They looked after a school experiment with two little spiders named Arabella and Anita.

In 1979, while Skylab was empty, its orbit became unstable and the world watched anxiously for weeks, wondering where it would crash and what would happen when it did. Most of the space station burnt in a spectacular fireworks display over Western Australia, but the more robust components fell to Earth over the towns of Balladonia, Kalgoorlie and Esperance. The pieces immediately became collectible: souvenirs of an experience most of us will never have. My fragment is part of the fuel tank insulation. It was taken to San Francisco when the *San Francisco Examiner* was offering a $10 000 prize to the first person who delivered a piece to their offices. There, it was further fragmented and divided up. I'm not sure, but this may be the same charred fuel tank that was displayed on the stage of the 1979 Miss Universe pageant held in Perth around the same time as Skylab's spectacular re-entry.

When I touch it, I think of the journey it's had and all the hands it's passed through. From Skylab's assembly at the Marshall Spaceflight Center in Huntsville, Alabama, launch from Cape Canaveral in Florida, into space, to Western Australia, then on to California, where a space fan acquired it and kindly gave it to me, back in Australia, nearly forty years later. This bit of space detritus has been treasured as a precious talisman in much the same way as a rabbit's foot or a piece of the Berlin Wall.

It's been on Earth far longer than it's been in space but somehow the scent of space still clings to it. This piece has found repose on my desk among other space souvenirs, like the mulga wood model rocket from Woomera and the grey metal Moon globe marked with the landing sites of early US and Soviet probes. Other fragments rest quietly in the vastness of the Western Australian landscape, slowly being buried by wind-blown dust, becoming an archaeological site which combines earth and space into a seamless whole.

CHAPTER 1

HOW I BECAME A
SPACE ARCHAEOLOGIST

And I am as far as an infinite alphabet
made from yellow stars and ice,
and you are as far as the nails of the dead man,
as far as a sailor can see at midnight
when he's drunk and the moon is an empty cup,
and I am as far as invention and you are as far as
    memory.

Susan Stewart, 'Yellow Stars and Ice'

As a child growing up on a wheat and sheep farm in the southern Riverina region of New South Wales, I was obsessed with the stars. The only other lights visible at night came from inside our own house. I'd go outside and look up, and wonder how the universe came to be. How to attain those other worlds? Why did we not have the technology to go and visit them? I longed desperately for the future.

And yet, the past was equally fascinating. Around me, on our property, were the traces of a multi-layered occupation: Aboriginal grinding stones now used as door-stops, trees with canoe scars standing in the middle of sheep pasture, abandoned wells dug by Chinese labourers, the farm machinery 'graveyard', the old pisé (rammed earth) home-stead now dissolving slowly back into the soil. Only in later years have some of the old farmers quietly mentioned to me the Aboriginal burial sites found throughout the region's sand hills.

In a long, circuitous journey, these two parts of my life came back together when I decided to apply archaeological principles to the stuff that humans have sent beyond Earth: the stuff we now call space junk.

OUTBACK AND OUT OF THIS WORLD

Our farm was not all livestock and crops. We also kept pigs, peacocks and racehorses. Racing was a Gorman family tradi-tion, although my father was not as serious about it as some of my cousins.

The cultural effects of the Space Race infiltrated even the world of country horse racing. During my childhood, our champion horse was called Gooyong. It would be nice if Gooyong (meaning 'camp') were a Wiradjuri word chosen to honour the Traditional Owners of our land, but the more likely truth is that my father found it in one of the many published vocabularies of Aboriginal words that were in circulation at the time. A popular one was by HM Cooper, published in 1948 as *Australian Aboriginal Words and Their Meanings*. Generally, the words were stripped of context, the differences between distinct languages elided as if Aboriginal culture was identical across the entire continent.

Gooyong, however, was not the original choice of name. She was born in 1957, the same year that Laika the dog went into orbit on the USSR satellite Sputnik 2, the second ever to successfully reach Earth orbit. My father wanted to call the foal 'Little Lemon', the English translation of one of Laika's nicknames. It seems an unusual choice for a farmer focused on wrestling a livelihood from the earth with the ever-present threat of drought. Perhaps the farm boy who rode a horse to school until he was eleven also dreamt of travelling to the stars. Too late to ask him now.

Laika's launch provoked worldwide protests about the iniquity of sending a non-consenting creature to die in space, as indeed she did. And this gives me a brief insight into my father's thought process. As with all farms, we had an array of dogs: sheep dogs, cattle dogs, hunting dogs, house dogs. I think Dad felt a bond with this far-away Soviet space dog and wanted to honour her in his own version of the Space Race.

The Board of Racing Australia, however, had other ideas. All horses have to be registered, and to avoid duplication of names, the board has to approve each one. Dad's application to call the new horse Little Lemon was rejected. He was very annoyed. But it seems that no-one else had applied for the name. Was this Cold War paranoia? After all, this was the country whose prime minister, Robert Menzies, refused to congratulate the USSR on Yuri Gagarin's first successful human spaceflight in 1961. Such politics didn't bother the Jockey Club of New York, who were happy to approve Sputnik, the first satellite launched in 1957, as an official racehorse name.

Funnily enough, Gooyong still had a space connection. Cooper's vocabulary was used to name streets in the space township of Woomera in South Australia when it was laid out in 1947, as a residential village for the employees of one of the first rocket ranges in the world. Gooyong Street is still there today. This provoked its own controversy, as a senior British military official thought the Aboriginal street names would make the job of postal delivery too difficult. He proposed using the American-style numeric system instead, for example, Fifth Street and so on. The Aussies were having none of it, and he was overruled.

I didn't know about Gooyong the horse's brush with space until many years later. But I remember going outside with Dad to gaze up at the night sky. The Milky Way arched over our house like a starry fleece flung in the dark light of the shearing shed. How could I not become fascinated by astronomy? There were constellation charts on the bottom shelf of the glass-fronted bookcase in the sitting room, and I learnt to locate and name them. The constellations were often figures from Greek and Roman mythology: Orion, Andromeda, Gemini, Perseus, and so on. I think the charts might have come from *Weekly Times* special offers, which was how we also acquired a *Reader's Digest* atlas. (The *Weekly Times* is a newspaper devoted to rural issues.)

THE MOON IN THE LIVING ROOM

I was one of 600 million people across the world who watched as Neil Armstrong set foot on the Moon in 1969. I was five years old, in my first year of school. It was a tiny two-teacher school in the nearby hamlet of Savernake. I say hamlet, because unusually for an Australian town, Savernake didn't even have a pub. It had burnt down in the 1930s and was never rebuilt. As well as the primary school, there were two churches, a post office, a community hall, and two houses – occupied by old Jimmy Crow the beekeeper, and the Millets, whose four daughters went to school with us. Jimmy Crow's bees made honey out of the noxious weed Paterson's Curse and the old-growth Yellow Box gum trees. As he got older his sight deteriorated, and the honey he delivered to my mother often had bits of burnt bee in it.

I was one of four in my class. On 21 July, we crowded into the teacher's residence attached to the school to watch the Apollo 11 Moon landing live on the television. (On the other side of the world, Erich Wendt was listening on the radio while he scraped back the layers in the Apollo 11 Cave.) There were about forty students from kindergarten to sixth grade. Mr Mitchell, the principal, arrayed us by height so that everyone could see the screen, and as the youngest, we four kindergarteners sat cross-legged on the floor right in front of the television. We had the best seats in the house.

What I remember is black and white blobs, the bulky body of the astronaut with the life support system on his back. It all seemed to happen in silence and slow motion. I don't remember words spoken, or the scratchy-beepy radio communications between astronauts and ground control, or any emotion other than excitement. I don't remember talking about it with the other kids, or having the principal explain what we were watching, although he must have done so. Just the slow majestic descent down the ladder of the man encased in a space-age sarcophagus. When it was over we went back to class.

This didn't make me want to be an astronaut, though. The Moon wasn't enough for me. Inspired by this and the young adult science fiction novels I later devoured, I imagined myself instead as the victorious commander of a massive intergalactic battle fleet. I had few scientific role models – apart from the wild-haired, lab-coat-wearing Professor Julius Sumner Miller on the television after school – so books became the door into the world of science. You can't be what you can't see, but you can read about it.

I suppose it was inevitable that I would become a bookworm. Both my mother and my grandmother, after whom I was named, read constantly. From the time I learnt to read, I did the same. Among the books in the glass-fronted bookcase were volumes of classical mythology retold for popular consumption, which I loved. The constellations united the stars and ancient civilisations in my mind. Perhaps it was not so illogical to love both astronomy and archaeology.

One day, sometime in the early 1970s, an encyclopaedia seller braved the long, winding track to our homestead, and found my father at home. No doubt there were cups of tea offered, maybe even a home brew: the result was a set of science and history encyclopaedias, complete with their own flat-pack bookcase. My father might have had one home brew too many when he assembled it, as it was always wonky and threatening to collapse.

Whenever I had time to read, I lost myself in those volumes. Dinosaurs, DNA and Dalí were entrancing to a country kid whose world consisted of sheep pastures and sand hills. Somewhere in the encyclopaedias I moved beyond dinosaurs and got stuck in archaeology, the study of the human past.

My grandmother's great friend Mercia not only shared her love of reading, but was a serious book collector. I was the lucky beneficiary of her passion for books. She gave me volumes of fairy tales illustrated by the greats: Edmund Dulac, Kay Nielsen, Arthur Rackham. She also gave me several books about archaeology written for children. They had a profound effect on me.

Chief among them was *Ancient Man: The Beginnings of Civilizations* (1922) by Hendrik Willem van Loon. Van Loon was a Dutch historian who migrated to America, and became a celebrated author of children's books. It was hard not to be impressed by someone whose name was actually VAN LOON. He did his own illustrations, which were kind of crap but also charming and quirky. Somehow he hit just the right balance between explaining things simply and inspiring you to want to know more. I was absolutely enthralled reading about the discovery of Neanderthals in 1829 and the Neolithic Swiss villages perched up on stilts above mountain lakes. I added *archaeologist* to the list of things I wanted to be when I grew up. A growing obsession with what lay beneath the soil did not diminish my stargazing proclivities, though. As evening fell I'd look for Venus, always the 'star light, star bright, first star I see tonight' to make a wish on. It was by far and away my favourite planet.

VENUS IN GLASSES

Among our milking cows were two beautiful Jerseys, with huge brown thick-lashed eyes, creamy fawn hides, and horns curved like the new Moon. Their names were Venus and Stardust (more evidence of my father's secret space thoughts, I now realise). My friend Ged remembers me pointing out Venus in the dawn sky when I stayed over at her parents' property, in our primary school days. Not only was the planet Venus the brightest thing in the firmament after the Moon, but Venus, the ancient Roman goddess of love, seemed much more interesting than any of the others – even the wise Athene (Minerva in the Roman tradition). And of course, Venus is the only female planet to identify with.

By now, I had read of how the ancient Babylonian astronomers charted the phases of Venus more than 2500 years before Galileo Galilei used his telescope to view the stars in 1610. Their observations of Venus were recorded in spiky cuneiform script on baked clay tablets excavated from the mound of Kuyunjik in Iraq, also known as the city of Nineveh. This early astronomy seemed all the more scientific and profound because I myself could not always clearly distinguish the stars from the planets. While the bright constellations were easy to pick out, much of the rest of the night sky was faintly blurred. For the life of me, I couldn't work out how the Babylonians had seen the phases of Venus, unless they had different eyes back then. (This may have been in one of my archaeology books, but sadly, while the Babylonians were fearsome astronomers, it doesn't seem to be the case that they recorded Venusian phases.)

I also could not understand how my father recognised species of bird from a distance (Neville W Cayley's *What Bird is That?* had the status of a bible in our house), assuming that some people could see things in patterns of flight that I was simply not capable of. I thought him terribly clever because he could identify a bird from its flying style.

But then came a revelation. In sixth grade, when I was ten, it came to my mother's attention that I was short-sighted, and probably had been the whole time. I'd just developed strategies to make up for the fact that I couldn't read the blackboard at school, and it never occurred to me that everyone didn't experience the world like this. So I was duly taken away to get glasses. Suddenly, my father's hitherto mysterious ability to identify flying birds made sense: he could actually *see* them. It was now much easier to tell stars and planets apart as the latter were no longer twinkling too, and I could at last see why people talked of the rabbit on the Moon.

However, this new sight was both a blessing and a curse. Glasses cemented a certain reputation captured by American writer Dorothy Parker's famous aphorism: 'Men seldom make passes at girls who wear glasses.' Being seen as smart was antithetical to being seen as desirable, the pinnacle of a woman's achievement. Athene was taking ascendency over Aphrodite as the contradictions and constraints of being a teenage girl were ushered in with high-school life.

ARCHAEOLOGY OR ASTROPHYSICS?

At primary school, I was a very literate child with a voracious appetite for fiction. Despite my love of science, I soon picked up the message that girls were supposed to be good at words and boys good at numbers. Girls also did sewing while the boys did woodwork, and girls were not allowed to compete in certain athletic events, such as the 400 metre run and the triple jump. It made no sense to the four girls in my class but we had a dim awareness that it was so we didn't damage our future fertility, whatever that meant. We grumbled at the unfairness of it and sneakily ran anyway.

Not yet twelve, I was sent to an impoverished Catholic boarding school in the nearest large town, Albury. It was not an edifying experience. Academic achievement was less important than religious devotion, and each day was a trial to be endured. Dormitory life was lacking in privacy, the food was execrable, and the nuns were casually vicious. I could tell you some stories; but just to give you an idea, in the early years we were woken every day at 6.45 am by a nun clanging a bell up and down the dormitory; we had to kneel down to say prayers before scuttling off to dress in our uniforms. Breakfast was stone-cold toast which we buttered and placed on top of our teacups so the steam would warm the toast and melt the butter. This had the side effect of little oily globules seeping into the tea below; buttered tea was the price we paid for the simulacrum of hot toast. You'd think it was the 19th century but this was the late 1970s – early 1980s. The terrible food and nutrition made many of us vow never to eat bad food again.

We were educated to be good Catholic wives and mothers. Nursing and teaching were considered highly suitable careers until we married and started reproducing. There was no particular emphasis on tertiary education: the nuns were more concerned that we didn't have sex before marriage or take contraception. (Hah! Shut up in the boarding house, I think I barely spoke to a boy who was not a relative or someone I grew up with until I was seventeen.) Sister Antonia lugubriously informed us of our duty to never deny our future husband his conjugal rights. It really was that bad.

I wasn't interested in nursing or teaching, as important as those professions are. I studied science all the way through high school, reading books about physics for my own enjoyment. Once we were allowed to use calculators from Year 10, I found that I excelled at mathematics. (How is it that school curricula take a subject so fascinating and manage to make it so dreary? Maths might be about numbers, but what you don't really learn at school is how intensely visual and spatial it is.) Despite my enthusiasm, I did not get a brilliant mark in physics. The common opinion of the adults around me was that my mark was insufficient to pursue a career in astrophysics. It was seen as more evidence that my strength lay in words, not numbers, despite high marks in chemistry and maths.

By then I was committed to archaeology, but I was still devastated to have the astrophysics option closed off. In the first year of my archaeology degree, I also took astronomy and persuaded the physics tutor at my residential college to let me audit the in-house tutorials. You could say I was 'leaning in' so far that I practically fell off the edge. But there was no-one to catch me. There was no-one who said, if that's what you really want to do, let's talk about how to achieve it. No-one said, hey, you're good at words but you're also good at numbers. In all likelihood I would still have gone down the archaeology path. But I don't know. It didn't seem as if I had that choice.

The kind of archaeologist I imagined myself being was a classical one, investigating ancient Greece and Mesopotamia. Or perhaps looking at the European past: tracking down the archaeology behind the myths of King Arthur or the fairy people of Ireland, the Tuatha dé Danann. Nothing I'd read or seen prior to university gave me the idea that one could practise archaeology in Australia – Aboriginal culture was regarded as a poor cousin of European and African archaeological records, on the periphery of where the real action was. Even now, people often express surprise when I say I'm an archaeologist working in Australia.

My views of Aboriginal culture were shaped by books in my grandmother's collection: Douglas Lockwood's *We the Aborigines*, Mary Durack's stories about Aboriginal children written for white children, Mrs Aeneas Gunn's *Little Black Princess of the Never-Never*, and the cartoons of Eric Jolliffe, which make for slightly uncomfortable reading now. I'd grown up on Aboriginal land surrounded by earth ovens, scarred trees and grinding stones, but my father broadly proclaimed that no Aboriginal people had lived there when white settlers came.

The perception that Aboriginal occupation lay lightly on the land was in the process of changing, partially thanks to chemist Willard Libby's method of radiocarbon dating developed in 1946, when he was working in US nuclear weapons research. In 1953, a cranium from a river terrace near Keilor, Victoria (Wurundjeri country), was radiocarbon dated to 14700 years ago. This was the first scientific evidence that Aboriginal people had been here for a lot longer than a few thousand years. Australian archaeologist John Mulvaney's 1962 excavation of Kenniff Cave in Queensland used radiocarbon to obtain a date of 19000 years ago, during the driest, coldest period of the Ice Ages in Australia (there wasn't that much ice on mainland Australia, just a drought lasting 20000 years). The following decade the date was pushed back even further. 'Mungo Man' was found eroding out of the interior of the dunes at the Willandra Lakes in NSW in 1974 and more than doubled this at 42000 bp. (The latest dates from the Northern Territory are over 65000 years ago. Archaeologists can no longer ignore Australia as an isolated backwater, and what happened here is coming to be seen as central to understanding the global dispersal of humans.)

By the time I got to university in the early 1980s, Australian archaeology was on a more respectable footing. My first year was an odd mix: astronomy, Ancient Greek theatre, ancient Mesopotamia, and Australian archaeology. We studied the river terraces of Keilor, and went on a site visit. The famous cranium was found in the Doutta Galla Silt layer. Doutta Galla, most likely the corruption of a Wurundjeri word, has a pleasing sound and I still like saying it. I think about Keilor every time I pass the Doutta Galla Hotel on the train on the way into Melbourne, although the hotel is nowhere near the site.

As the next few years unrolled I wrestled with the impossibly complex verbs of Ancient Greek, abandoned Plato in favour of Aristotle, and discovered the 'New Archeology' theories of the 1960s – a bit late, but oh well. I completely pissed off a lecturer by borrowing the textbook from him the day before the exam and then getting a high mark in it. Exams were held in the huge, echoey Exhibition Buildings in Carlton, now on the World Heritage List. In the holidays, I worked as a kitchen hand to supplement my meagre government funding. And I found something unexpected. I didn't love the archaeology of the great northern hemisphere civilisations as much as I thought I would. I didn't want to be a classical archaeologist. In truth, it wasn't scientific enough for my liking.

BACK TO THE PAST

Armed with an Honours degree, I found myself looking for work in Sydney. A major source of employment for archaeologists outside the academy was heritage consulting. Cultural heritage comprises the objects, places and practices from the past that people think are important in the present and want to preserve for the future. It's about community identity and well-being; feeling connected in an age of globalisation and social fragmentation.

From the 1970s, each state has had legislation aimed at protecting Indigenous heritage from destruction by development, and archaeologists were brought in to identify and assess the cultural significance of places threatened by urban expansion, mining, roads and other infrastructure. This work was done (ideally) in collaboration with Aboriginal communities. It was a fraught space to work in. The disjunction between a heritage relegated to the past and the living communities grappling with dispossession and discrimination, the legacy of the White Australia policy, and the government's assimilation policy of the 1960s, was played out in paddocks and bushland where Aboriginal heritage was being erased.

Our task was to identify heritage sites, which might be culturally modified trees, concentrations of stone tools, rock art, axe-grinding grooves, quarries, burial sites, hearths, fish traps or stone arrangements. Often, we were contributing the heritage component of an environmental impact statement. In theory the legislation protected any site of significance to the Aboriginal community or to scientific investigation, but in reality, when it came down to heritage versus development, development usually won.

One job led to another, and soon I was living the life of an itinerant archaeologist, participating in surveys and excavations around Sydney's Cumberland Plain and the coal mines of the Hunter Valley. This was roughly how the process worked. The client – a mining, construction or environment and planning company – would contract a heritage consultant to make sure they did not commit an offence under the relevant heritage Act by damaging or destroying Aboriginal heritage. The heritage consultant's job was to firstly identify what heritage was there, through research and looking at registers of known sites. You could assess the potential for sites to exist by looking at the landform and geology. Was there an old eucalypt forest on the riverbank? This might be where you'd find scarred or culturally modified trees, where bark had been removed to make coolamons, canoes, or shelter construction materials, or footholds cut to ascend the tree and locate a beehive to extract the honey. Given the extensive land clearing of European agriculture, these trees diminished in number by the year. We were trained in how to distinguish bark removal by humans from branch fall, lightning strike and other natural causes of tree scarring. Were there sandstone escarpments? They were likely to be full of rockshelters where you might find art or deep deposits full of hearths and stone tools. You could date the layers, particularly if there was charcoal in them. Did the geological map show outcrops or crusts of silcrete, a brittle, highly siliceous rock that was perfect for making stone tools? Then it was likely you would find a quarry site there, where Aboriginal people came for stone-working supplies.

The next stage was a survey on the ground. (Archaeology involves far less excavation than you might think.) The archaeologists would walk over the area where the mine pit, or the road, or whatever, was going to be built, and look for the sites we knew about and the ones we didn't yet. The sites would be recorded in full detail, as they might not survive the development construction. We would do a significance assessment for each one using the criteria in the Burra Charter. This was a set of guidelines for working out what kind of cultural significance, and how much, any place or object might have, first developed by the Australian Chapter of the International Council on Monuments and Sites in 1976. The Charter guidelines are so clear and straightforward to apply that they are used all over the world.

First, you look at historic significance: association with a historical event, person or process. Scientific significance is about research potential. What could we learn from this site or object? If it is destroyed or damaged, what questions might we no longer be able to ask or answer in the future? Aesthetic significance was about sensory engagement: size, scale, sound, touch, taste, smell, visual appearance. Social significance was an indicator of community esteem, the emotions and attachments people felt towards a place. Part of that might be spiritual significance. Usually there are aspects of all of these, but you might find a site that had high historic and scientific significance that no-one cared about. Equally, a site might not appear to be anything special, but would have extremely high social significance for the community attached to it. The heritage consultant's job was to make the case to the developer for why a place should or should not be protected.

Ironies abounded in the fact that usually white, university-educated archaeologists were charged with making decisions about someone else's heritage, wrapping it up in scientific terms to meet the legislative requirements. My Aboriginal colleagues on these jobs were usually less interested in the archaeological theories about the sites, and more concerned with how they could be used to maintain culture and connections to country in the present. This was a lesson that the past did not rest easy. All the same there was a sense of a common goal, working together to prevent the complete erasure of the physical traces of intangible culture from the land. It was the social values of heritage that were the most significant; it was no good focusing on the past if it did not challenge the present.

STORIES FROM STONE

This work sparked my interest in stone tools. Some excavations produced thousands of stone tools from sites soon to be swallowed up in an open-cut mine pit. In the Hunter Valley, there were glossy silcretes of rose and champagne pink, dusty mustard-yellow mudstones and brilliant white quartzes. While, to the casual observer, they might look like broken rocks, to the trained eye there were angles and features that betrayed the deliberate intent of human hands, and an insight to how people lived in this landscape thousands of years ago.

And also, sometimes, very recently. Occasionally, there was dark green bottle glass that had been deliberately flaked. It was much more difficult to distinguish this from natural breakage, but I learnt how to do it. This was evidence of how Aboriginal people had adapted to a new material introduced by the European invaders. Glass behaved just like a highly siliceous and brittle stone and produced an exceptionally sharp edge.

After a few years, I transitioned from 'dig bum' to stone tool analyst. But I thought the stories told about these arte-facts were repetitive and often lacking in imagination. The beautifully manufactured 'backed artefacts', small blades with tiny flakes removed to make a steep margin, were always proclaimed to be spear tips and hence evidence of hunting. Hunting was evidence of meat-eating, and meat-eating was thought to be the catalyst for brain growth and bipedalism back before hominins left Africa. Since stone tools were the only technology that survived archaeologically for millions of years and across several hominin species, it was assumed that they were male technology. It said so on the box: man the toolmaker, man the hunter. Women gave birth, cowered in the backs of caves, posed as the model for a Venus figurine occasionally so that Palaeolithic 'man' could get his other rocks off, and maybe collected a worthless vegetable from time to time when the mammoth chops were running low. The sometimes openly stated and mostly implicit assumption was that human physical and cultural evolution was driven by male hunting. Was this the best we could do?

A trip to the UK provided the catalyst to turn these thoughts into a more substantial project. In the Pitt Rivers Museum in Oxford, I found a set of flaked bottle glass tools used for body modification, made by women from the Andaman Islands, part of an archipelago in the Indian Ocean, in the 1870s. They had been collected by Edward Horace Man, who ran a 'home' for the Andamanese after British settlement started to encroach on their land and erode their society. Men didn't use stone tools in this period of Andamanese history, as stone and cutting activities were the realm of women. If they had, it's likely the colonial administrators would not have even noticed what the women were doing. At first I was interested in the technology transfer from stone to glass, but soon I realised there was something else far more interesting going on.

Andamanese women of the Aka-Bea and other groups started using bottle glass instead of the traditional quartz to carry out scarification and hair maintenance. Scarification is related to tattoo: scars are cut into the skin to produce a raised pattern, which is both visual and tactile. Often, the scars were first cut as part of an initiation ceremony. In the Andaman Islands, scarification was part of a complex cosmology in which height above the surface of Earth, and the physical density of objects, were related to smell within the thick forests of the islands. Scarification wasn't just about adorning the body: it was a mechanism for demarcating space and time. Yet, despite the central importance of this practice for constituting the social body of an Andaman Islander, the glass and quartz flakes were nothing special. They looked quickly made and quickly discarded, the sort of thing that archaeologists would call 'waste' and pay no more attention to, while they were focusing on the fancy hunting tools everyone seemed so obsessed by. It made me aware that what looked like everyday tools could in fact be related to highly nuanced cultural practices.

These artefacts became the centrepiece of my PhD at the University of New England, which I started in 1994. I realised that if I could identify a pattern of residues and use marks which indicated a stone tool had been used for body modification, then this could be used to search for the earliest evidence of symbolic behaviour – and hence the emergence of behavioural modernity. I spent a year peering through a microscope at the Andamanese glass tool edges to determine if there was a consistent signature of skin, hair, oil residues, and faint polish and scratches. Happily for me, there was. Potentially, you could use this method to demonstrate that people had been modifying their bodies hundreds of thousands of years ago. But after finishing that work in 2000, I never wanted to look down another microscope ever again. I was broke, unemployed, and tired of living the life of an impoverished student. I took the first job I was offered, in Queensland. This was the setting for what happened next.

LYING IN THE GUTTER, LOOKING UP AT THE STARS

My new job was Project Archaeologist for the raising of the Awoonga Dam, near the central Queensland town of Gladstone. I rented a lovely old Queenslander house. The house had the characteristic broad verandahs of that architectural style and a back garden with guavas, mangos, poincianas, and other marvellous semi-tropical trees. It also had an excellent bath, fabulous for soaking off the dirt after a hard day in the field.

I was frequently in the field with my team, all of them women from the three Native Title claim groups in the area. Surveying, monitoring earthworks, excavating, a whole bunch of stuff. In the height of summer, it could be very hot and sweaty work indeed. On one such day I came home, exhausted, clumped up the stairs in my steel-capped, acid-resistant boots, flung off my fluoro vest and hard hat as I entered the door, and went straight to the fridge for a delicious cold beer.

Now I have to confess I am slightly on the old-fashioned side in adhering to the principle of changing for dinner, whether one is at home, or in the field with only a flimsy dress which has been rolled into a ball and squeezed into some corner of the suitcase not occupied by Explorer socks. But sometimes it is just too much of an effort, and this was one of those days. The next stop after the fridge was the verandah, where I collapsed into a chair with my beer and sat, thinking of nothing much, looking up at the stars.

Queensland doesn't have daylight saving, so it gets dark very quickly on summer evenings. The stars were already out and Venus was bright among them. I gazed up at the dark of space punctuated by luminous worlds, thinking about the Babylonian astronomers whose feats of observation so puzzled me as a child, and the constellation charts that were the keys to making meaning of the impossibly distant stars. The constellations were imaginary lines drawn between stars which had little basis in reality, but which turned the night sky into a place where you could read signs and symbols. Then something occurred to me: the bright points of light are not just stars. The sky is full of satellites and space junk too. There are artefacts in space, made by human cultures, and discarded in the same way as stone tools across the landscape. What lines could we draw between them to make meaning from this modern junk pile?

It was the second part of the thought that was critical, very much related to my task of managing the heritage values of more than 300 recorded Aboriginal and European sites within the inundation area of the Awoonga Dam. If there is human material culture in space, does it have heritage value? Does the Burra Charter apply to things that aren't even on Earth? I thought about this for a while, my beer forgotten. I decided I was going to find out.

It wasn't that no-one had thought of it before. In the 1970s, American historical archaeologist James Deetz said that one day we might be doing an archaeology of starships. William Rathje, famous for analysing the garbage of modern households in Tucson, Arizona, had written about the archaeological potential of orbital debris in 1999; and in New Mexico, Beth Laura O'Leary was making a catalogue of all the Apollo 11 artefacts left behind on the Moon. Greg Fewer in Ireland had suggested a heritage listing process for sites on the Moon and Mars in 2002. There was already a context for the archaeology of space. But for me at that moment, it felt as if something had clicked into place. My childhood connection with the heavens, which I'd broken by pursuing what lay beneath the earth, was forged again.

Later that year (2002), I finished the Awoonga Dam contract and took up a position as Senior Conservation Officer in the Heritage Branch of the Environmental Protection Agency (EPA) in the Central Queensland town of Rockhampton, the 'beef capital' of Australia. I moved from Gladstone to Rockhampton's nearby town of Yeppoon. After work and on weekends, I'd sit at my tiny desk with a dial-up internet connection and surf the web to make up for lost time. I made tables of spacecraft and launch dates; I read histories of space programs and trawled through sites like *Encyclopedia Astronautica*. And I became aware, for the first time, that Australia had a rich and diverse space history of its own. Australia had, in 1967, been the third (or possibly fourth) nation to launch a satellite from within its own borders! I was astonished by how little I knew of events that took place during my childhood.

In these early days, themes that have since become dominant in my work started to emerge. The Space Race and the triumph of the white, male American astronaut seemed an inadequate lens through which to view the development of space exploration, just like 'Man the Hunter' didn't really catch the complexity of human behaviour in the past for me. I was more interested in how everyday people engaged with space through projects like Vanguard 1, the oldest satellite still in orbit, and Australis Oscar 5, Australia's first amateur satellite in 1970. Both of these satellites relied on amateurs and enthusiasts – what we'd now call citizen or community scientists – to assist with tracking them and collecting data. This was more like the sort of story I wanted to tell. With my background in Aboriginal archaeology, it also seemed obvious to investigate how Indigenous people had experienced the Space Age, at places like the Woomera rocket launch site in South Australia. I thought about what an archaeological approach could bring to the study of space exploration. How would you apply Burra Charter principles to stuff in space?

I had to work out what my own parameters were too. I already had a reputation for tackling quirky research questions after my PhD on body modification, and I needed to make sure that my approach was rigorous. The first thing that anyone would think of would be 'aliens'. SETI (the Search for Extraterrestrial Intelligence) is a legitimate field of research, and I was as fascinated by it as the next person, but I couldn't afford to sabotage my version of space archaeology before it started. I decided to stay right away from astrobiology, the investigation of life on other planets, and SETI (I'm not so rigid about this now). I also decided to rule out astronomical heritage. Here there was a huge overlap with space science, so it was impossible to disentangle them to a degree, but one couldn't do everything. My focus was going to be satellites and space junk.

LAUNCHING INTO ORBIT

Now a more serious issue raised its head. Already, I was struggling to access the information I needed with the incarnation of the World Wide Web current at the time, and in the library of Central Queensland University. I was doing this in my spare time, and it wasn't easy to negotiate leave to attend conferences that weren't strictly work-related. I would be taking leave to attend the World Archaeological Congress in April 2003, where I presented two papers in the first ever conference session on space archaeology. When I returned, nothing would have changed. How serious was I? I searched my soul and decided that I had never been more serious about anything in my life. So if this was the case, I had to accept that living in Central Queensland and working at an unrelated job was not going to further my research. I was going to have to leave.

This was a tough realisation, because I loved my life at that point. Working at the EPA was great; I had excellent colleagues, and I enjoyed the pace of the public service. I loved living in the beautiful coastal town of Yeppoon. I was presenting a music show on the local community radio station, Radio 4NAG, and my radio friends were a delight and joy. Every now and then I would escape to Brisbane for a touch of city life. It was all pretty excellent. But to make my vision of space archaeology happen, I would have to quit my job, leave Yeppoon, and take a leap into the unknown.

I submitted my resignation, effective from the end of March 2003. I packed up my house and put everything into storage, keeping only one suitcase, my laptop, and some books and papers. It would be seven years before I was able to retrieve my possessions. But that's another story.

From that time, I became a space archaeologist. I looked at every planet and moon in the solar system from the perspective of the human artefacts in its orbit or on its surface. I gazed at Venus, my landmark of dusk and dawn, no longer puzzled by the prowess of the Babylonian astronomers, but full of wonder at the Soviet spacecraft stranded there.

One day, many years later, an astrophysicist of my acquaintance said, 'Can I ask you something? I don't understand why, if you're so into space, you didn't just study astronomy or astrophysics at university'. It was a good question, and one that no-one else had ever asked me before.

'I did in fact want to be an astrophysicist. But I didn't get a good enough mark in physics,' I replied. 'Why? Did you fail physics?' 'No, I didn't. I passed it.' I had not thought of it like this before. It was true, I hadn't failed physics.

'If you don't mind me asking,' he said, 'do you remember what mark you did get?' Yes, I did remember. It was burnt in my brain, the not-good-enough mark that also said I wasn't good enough. Feeling slightly hesitant and a little ashamed, I revealed it.

'Really?' he said. 'Do you know, that is the same mark I got for HSC physics.' The world turned slowly upside down and then back the right way up. We got the same mark. He went on to get a PhD in astrophysics. I was encouraged to stay with words, through the weight of tradition that held a woman had to be outstanding to succeed in a non-traditional field whereas a bloke just had to be regular.

I'd spent the intervening years looking down, not looking up. But you could really think of it as two sides of the same coin. The old technologies on Earth were the precursors of those which enabled celestial travel, and this more recent phase of human existence was as much about symbolic behaviour as anything that went before.

A few years ago, the first edition of the *Encyclopedia of Global Archaeology* was published, complete with entries on space archaeology and heritage. Travelling encyclopaedia sellers probably don't roam the countryside any more. But if they did, perhaps they'd have a yarn with someone like my father, over a casual beverage, and say, 'Look at this one! It even has space archaeology in it!'

CHAPTER 2

JOURNEY INTO SPACE

Our trajectory
the nautilus shell curved path
away from our home.

Christine Rueter, 'Our trajectory'

Space exploration emerged as part of the Cold War conflict between the USSR, representing communism, and the
US, representing capitalism. For a child growing up in the
1960s and 1970s in Australia, the Cold War didn't seem so
cold. The Vietnam War (1962–75) was in the news every day
and seemed interminable. I wondered if it would ever end.
People talked of the Domino Effect: like a row of domino
game pieces tumbling down one after another, if south-east
Asia fell then Australia would be next. This fire was fanned
vigorously by the conservative Catholic journalist and social

activist BA Santamaria, who was on the television every Friday night inveighing against the dangers of communism and the 'permissive society'. (Incidentally, his daughter Mary-Helen Woods, a superb musician, was my violin teacher.)

My father was as reds-under-the-beds as they came and was convinced that we were being indoctrinated with communism in school. I did nothing to allay his fears with my nascent feminist consciousness. It was Cold War for sure in the Gorman household, as Mum encouraged us to put our spare change in the Project Compassion money box stationed on the kitchen bench during Lent to raise money for starving children, and Dad tried to persuade us not to, as some of those children might be communist. For the record, we ignored Dad on this one and the cardboard box was stuffed with one- and two-cent coins by the time Easter arrived and we delivered the money to the church in Savernake. I guess this was the way the Cold War played out in daily life.

The competition between the US and USSR to achieve not only military, but also ideological supremacy led to what has been called the 'Space Race': a race to be the first in orbit and the first to the Moon. A compelling story of daring courage and creative science, it's the background to the modern world's reliance on satellite technology. This is also the period when the planets ceased to be objects observed from far away and became more like neighbours. Deep space probes ventured out into the solar system and sent back images of landscapes very different from Earth, but also familiar, evoking the deserts and icefields of our own planet.

I've always felt slightly uncomfortable with the Space Race narrative. It's too triumphalist, too rooted in conflict. It promotes the idea that technology is driven by competition and we just have to put up with a little war on Earth to get into space. This isn't the world I want to live in. As I became aware that Australia had its own history of space exploration, I began to seek out other forgotten stories of space, the ones that didn't always make it into the standard histories or television documentaries. I felt the stories that didn't fit neatly into the Space Race museum displays emphasised different qualities and different communities, ones that interested me much more.

Of course it's not possible to ignore entirely the role of the original spacefaring nations and the political context of space exploration; we need historical background for understanding where we want to go in the future. So I'm going to approach this from a more archaeological perspective, by using objects rather than events to weave a web of connections. The questions I want to ask of these space objects are those of the Burra Charter and its categories of cultural significance. Who cares about them – who are their communities? How do they engage our senses? What technology do they employ? What vision of space or the universe do they represent? What values do they promote? What future do they predict? For each decade from the Second World War, I'll focus on a place or object that is relevant to these questions, that represents the stories which captured my attention as I roamed through the launching pads of space history.

1940s: A ROCKET AND A BOMB

Every archaeologist knows that it's a hopeless task to pin-point the first time a technology was used – like the first wheel-thrown pot, or the first blue cheese. We see processes at a broad scale, which makes such a quest meaningless. However, this all changes for the historic period when we have documentary records and access to the people who were there. And contemporary space exploration really did start from one artefact. This is the V2 rocket, the German inter-continental ballistic missile which became the first capable of reaching outer space, in 1944. The V stands for *Vergeltungswaffe*, or Vengeance Weapon. From this geographic and chronological point, rocket scientists and rocket technology radiated outwards to other nations and other cultures, like a stone dropped into the smooth surface of a pond.

What did this rocket look like? It was 14 metres high, 1.65 metres in diameter, and weighed 4000 kilograms when empty of fuel. The body was a gently curved cylinder, with four fins at the base, and the top tapering to a point. This is the archetypal rocket, reproduced in toys, animations, T-shirts, models – whenever a symbol is needed to evoke space travel. It's the shape that signifies 'rocket', whereas before it might also have signified 'fish'.

From a distance the exterior looks smooth; but when you get closer, you can see that it's made out of panes of metal bolted together. The bolts sit in neat, evenly spaced rows along the seams, much as you might see if you look out of an aeroplane window over the wing. In fact, they are very similar: the aeroplane too has a cylindrical body with a tapering nosecone and fins – although it flies horizontally rather than vertically.

The exterior design and colours varied. For test rockets, the body was painted in a black-and-white chessboard pattern, so that as it spun around its long axis, the spins (known as the roll) could be measured. The V2s that were sent out on their missions of destruction were painted in a camouflage pattern of jagged green, brown and white shapes or just plain olive green.

The rocket produced sounds, although not immediately. This was very different from its precursor, the V1 'rocket bomb' or doodlebug, which had such a distinctive buzzing sound it became known as the buzz-bomb. The V2 travelled four times faster than sound, at 5000 kilometres per hour at its fastest, so it struck without warning – only after it had landed did you hear the noise. People neither saw nor heard them approach. The first hint of the rocket was the explosion as it struck the earth with its warhead, taking down buildings which splintered into bricks and broken glass. Then the sonic boom hit, and sometimes the roar of the rocket followed it to ground. This is how Thomas Pynchon describes it in the novel *Gravity's Rainbow*:

> But a rocket has suddenly struck. A terrific blast quite
> close beyond the village: the entire fabric of the air,
> the time, is changed – the casement window blown
> inward, rebounding with a wood squeak to slam again
> as all the house still shudders.

The rocket has a particular movement, too. It's a ballistic missile. You point it in the right direction, give it some power to blast off and to guide it for a short distance, and then let gravity and speed take over. To reach another continent, the rocket ascends to sub-orbital heights, but without more fuel and power it can't break free of gravity and so begins its descent back into the atmosphere to impact. It's a smooth elongated rainbow of a curve with the scaffolding of calculus supporting it.

The Second World War was the catalyst to develop a rocket capable of crossing the continents to deliver bombs – and, later, reach space. The V2 is strongly associated with one man, the German rocket scientist Wernher von Braun, and countless unnamed people who died in order to make it. Around 95 per cent of the 6084 V2 missiles that were man-ufactured were constructed by 20 000 slave labourers from concentration camps in the last seven months of the war. Of these, 12 000 died in the process from starvation, overwork and cruel treatment.

Armed with a one-ton warhead, it was a weapon aimed at civilians, and its intent was to produce fear as much as death. The V2 rocket rained destruction and havoc on London, Belgium and France from 1944 to 1945, leaving landscapes pocked with craters and strewn with wreckage. The craters were typically 18 metres wide and 5 metres deep, and the impact of the rocket caused 3000 tons of material to be ejected out of the earth. Nine thousand people died in the UK and Europe as a result of V2 strikes.

After the US dropped atomic bombs on Nagasaki and

Hiroshima in 1945, a terrible offspring of these two technologies was conceived: missiles that could deliver nuclear weapons from the edge of space. In the last days of the war, as they advanced through the occupied countries of Europe, the four allied nations of the US, USSR, France and Britain competed with each other to obtain both the rockets and the rocket scientists, chief among them Wernher von Braun. These four states became the first spacefaring nations thanks to a technology of war.

Only eighteen of the 6000 or so V2s made remain intact. Two are in the collections of the Australian War Memorial (AWM). In 1957, one of these rockets went on a road trip around Australia together with other German and Allied missiles, as they were transported from the Woomera rocket range to the AWM in Canberra. A series of photographs in the AWM shows the rocket on its original launch trailer driving through suburban streets in a convoy of other decommissioned weapons. One image was taken in Macleay Street in the Sydney suburb of Potts Point, just round the corner from where I lived in the early 1990s. If I'd lived there in the 1950s, I think I'd have been out in the street watching as the missiles passed by the elegant Macleay Street Regis apartment building (it was my ambition to be able to afford to live there), wondering what was happening out in the Woomera desert now that they'd gone.

The convoy stopped in Martin Place, in the centre of Sydney, where crowds flocked to see the famous weapons. A cut-out was made in the body of the V2 so that the internal mechanisms could be viewed. This particular V2 had begun its transition from weapon of war to historic artefact. It came to a final rest as a museum artefact in the same year that its descendants realised the dream of reaching orbit.

1950s: WAGING PEACE IN THE COLD WAR

By the early 1950s, rocket technology built on the V2 had developed to the point where the first satellite launch seemed possible. The global scientific community had been working towards a massive co-operative effort to study Earth, called the International Geophysical Year (IGY), to take place in 1957–58. What could be better than measuring Earth from the outside? Everything we knew about the space environment at that time, we had learnt from inside the envelope of the atmosphere. The first satellite could change everything. This was before NASA existed, and the United Nations space treaties had not yet been written. The IGY was effectively building the first road map for using space.

The oldest human object in space, our first piece of genuine space archaeology, is the US satellite Vanguard 1, launched as part of the IGY program. It's a polished sphere of aluminium just 16 centimetres in diameter, weighing 1.47 kilograms. For comparison, a standard basketball is 23 centimetres and weighs a bit over half a kilogram. Compared to Sputnik 1, the first satellite to reach Earth orbit in 1957, it was tiny – the size of a grapefruit, as Soviet president Nikita Khrushchev joked somewhat nastily. You could easily pick it up and throw it in a backyard game. But it is no featureless ball. On the outside are six 5-centimetre-square solar cells, as well as six antennas oriented at right angles to each other. The antennas pass through the body of the satellite, crossing inside. The solar cells were divided into two sections and looked a little like window frames stuck on the surface, or perhaps cucumber frames. There's a seam around the middle which joins the two hemispheres together. The satellite described an elliptical orbit around Earth, 600 kilometres at its closest, and 3000 kilometres at its furthest. It spun and rotated as it hurtled through space, endlessly falling towards Earth.

Both Sputnik 1 and Vanguard 1 were aluminium spheres, little artificial moons that circled close to Earth under the watchful gaze of the real Moon. The French called this design 'bébé lune' or baby moon, as if it might grow up into a big one, or was some kind of a moon seed. The silver surface was polished to make it highly reflective: these satellites were meant to be seen.

And heard. Vanguard 1 transmitted a radio signal back to Earth for six years after its launch. The signal has been recorded; listening to a snippet, it sounds like a theremin chugging back and forth on a train track. For scientists and ham radio operators listening back on Earth, the aural signatures of the early spacecraft were as distinct as their appearance.

Vanguard 1 was intended to make the US the first nation in space. The term means 'leading the way' but also refers to the advance troops of a military attack. Space exploration was not just about science; it was also about winning hearts and minds. These first satellites were ideological weapons to demonstrate the technological superiority of capitalism – or communism.

The problem was that the IGY was a civilian scientific program, but the rocket programs were military. How to present a US satellite as a weapon of peace rather than war? Vanguard 1 was the project of the US Naval Research Laboratory (NRL). They needed to give the satellite a civilian spin to present the US's intentions in space as peaceful. This meant the launch rocket should not be a missile, but a scientific rocket, made for research purposes. Such sounding rockets were, however, part of the military programs too – their purpose was to gather information about the little-known upper atmosphere for weapons development.

Unfortunately, Project Vanguard was beset by problems. Sputnik 1 beat it into orbit when it leapt off Earth on 4 October 1957 and sent shock waves through the world with its distinctive beep, especially in the US. People didn't really understand what this meant: was it a weapon, would it be dropping bombs onto Earth from orbit? But nothing exploded, and soon the lonely satellite was joined by others: Sputnik 2 with Laika the dog on board in November 1957, and the US Explorer 1 on 31 January 1958. Project Vanguard continued to struggle, with a spectacular explosion just after lift-off during its third test in December 1957, which earned it the nicknames 'Flopnik' and 'Kaputnik'.

Finally, Vanguard 1 was successfully launched on St Patrick's Day, 1958. For a short time, between 17 March and 14 April, there were three of the four earliest satellites orbiting at once, and they couldn't have been more different. Sputnik 2 carried the corpse of Laika, the first death in space, encased in her Dalek-like conical capsule still attached to the core of the 270-ton R7 rocket which launched her. The US Explorer 1 was a small package of scientific instruments inside a cylindrical stainless steel casing painted with white stripes. It had four whip-like antennas attached to its body. It was also attached to the fourth stage of the launching rocket. And then there was Vanguard 1. Put them on a slide under a microscope and you could be looking at plankton: the same phylum, but different species.

The little grapefruit made its presence felt far across Earth. The astronomer Fred Whipple, from the Smithsonian Astrophysical Observatory, had an idea for the IGY satellite program that would help Project Vanguard present the right image and contribute to the scientific outcomes. It was all well and good to launch a satellite, but you also had to know where it was in space so that you could collect its data. In the 1950s, the technology to do this was still in its infancy. And in the words of science fiction author Douglas Adams, space is big. Really big. When something the size of a grapefruit is launched, you can predict where it should end up, but you don't know if it's there until you've seen it. Someone has to look for it.

This was the purpose of Whipple's Project Moonwatch. Volunteers – nowadays we would call them citizen or community scientists – across the globe watched for the satellite using binoculars and telescopes supplied by the Smithsonian. There were more than one hundred teams, including several in Australia. They divided the sky into small overlapping portions and each person was assigned one portion to scan.

As well as the volunteers, several nations hosted a unique type of antenna to pick up the radio signals from Vanguard 1. We're very familiar with dish antennas today, but the earliest tracking antennas were of very different designs. The Minitrack interferometer system created for Vanguard included antennas which looked like metal railway tracks up on stumps. They were located in four countries, including one at Woomera in Australia. Vanguard 1 might have been launched by the US, but its success was international.

Vanguard 1 has become significant by lasting the distance. In 2018, it was sixty years old. Its high orbit protects it from being pulled back into the atmosphere in the short term. We know where it is, because its position is tracked; but it may not be as shiny as it once was, after being bombarded by micrometeorites and millions of tiny specks from the decayed space junk swirling all around it. A recent study looked at the reflectance signature of spacecraft surfaces when you bounced light off them, and found that those which have been in orbit the longest appeared redder, indicating rougher surfaces. We may presume that Vanguard's polish has faded, although the sphere is intact. Perhaps it is pitted like asphalt or a miniature cratered moon. It's the only object that can tell us what happens when a human artefact is exposed to the space environment for that long.

The historians Constance Green and Milton Lomask say that Vanguard 1 is 'the progenitor of all American space exploration today'. The little satellite meant to represent the peaceful uses of outer space is a physical reminder of the competition to imprint space with meaning in the early years of the Space Age.

1960s: ... AND ALL I GOT WAS THIS LOUSY DUST

Less than a decade later, both the US and the USSR had their sights set on the Moon. By 1969, when the Apollo 11 crew set off, there had already been twenty-six spacecraft from both the US and USSR which had crashed or soft landed on the lunar surface, a substantial archaeological record. Among all the landers, rovers, probes and flags left on the Moon, however, is a tiny matchbox-sized instrument that seemed insignificant at the time.

The Dust Detector Experiment (DDE) weighed just 270 grams and cost about $50 to make. It consisted of three solar cells, two on the sides, and one on the top. On the back of each cell was a thermometer, and a circuit board wired up to measure voltage. If dust landed on the solar cells, they collected less light, which caused a change in voltage. The aim was to measure how quickly dust accumulated. The DDE was mounted on a 47-kilogram instrument called the Passive Seismic Experiment (PSE), meant to measure surface disturbances. It was part of the Early Apollo Scientific Experiments Package (EASEP) deployed by the Apollo 11 crew.

The minute I learnt that lunar dust might be arranged in airy microscopic towers called fairy castles, I was hooked.

Lunar dust has been formed over billions of years by the bombardment of meteors that break down the bedrock (or regolith). The dust is incredibly fine and electrically charged. The fairy castle structure is one way of explaining the unusual way the lunar surface reflects light, so they are only hypothetical; but their existence would be hard to ground-truth as a footprint or the blast from rocket exhaust would crush the castles flat.

Have you ever driven along a country dirt track deep in 'bulldust', incredibly fine dust that sneaks through the window seals and makes you cough? It rises in clouds around vehicles and settles in drifts on the windscreen and in any available nook and cranny. This is what lunar dust is like, only much much finer and stickier.

After Neil Armstrong descended the ladder and said his famous lines, he and Buzz Aldrin had a lot to do. They had to assess their environment: was it safe to move about? There were ceremonial activities, the phone call to the White House, and the raising of the flag, which meant they had to have the TV camera set up first. In between, the astronauts went about collecting samples and setting up experiments. The PSE and the Lunar Laser Retroreflector, a block of glass prisms shaped like the corner of a cube designed to reflect laser beams from Earth, were the last to be set out, around 4.30 pm on the afternoon of 21 July (GMT). These experiments would keep on working after the astronauts left. The PSE was placed 17 metres away from the lander and ascent module. The DDE was so small that you could easily overlook its presence on the insect-like PSE, with its proboscis antenna and solar panel wings. The PSE and DDE sat on top of the lunar reg-olith – there was no weather, animals or curious humans to interfere with the experiments – and waited for something to happen. A moonquake, a meteorite impact, or the flurry of dust when the ascent module lifted off. Their wait was silent and still. Perhaps the experiments would have made tiny electronic sounds, but there was no atmosphere to convey the soundwaves and no ears to hear them – the astronauts were cut off from the environment by their white suits.

The first serious dusting of the DDE surfaces might have been when the astronauts started jettisoning stuff they didn't need, throwing it out of the hatch of the ascent module, to make sure it was light enough to take off with its new cargo of Moon rocks. Among the stuff they discarded were the life-support packs, arm rests and scientific equipment. The layer of dust this spread on the DDE solar cells would have been overlaid as soon as the ascent module started its engines, sending higher energy plumes of dust over the surface.

The DDE was the brainchild of Australian scientist Brian O'Brien. In the 1960s, he was professor of space science at Rice University in Houston, Texas. NASA put out a call for experiments for the Apollo missions in 1965, and O'Brien proposed an experiment called the CPLEE (Charged Particle Lunar Environment Experiment) to measure radiation, which he'd already successfully tested on his own small satellite Aurora 1. He conceived the dust detector as a way of managing the risk to the CPLEE from dust. It was really an afterthought. O'Brien's CPLEE was one of only seven experiments chosen from ninety proposals, so the dust detector piggy-backed on a ride to the Moon as the eighth experiment. The same device was deployed by the Apollo 12, 14 and 15 missions. Although there were concerns about lunar dust, no-one yet knew that dust was going to become one of the most troublesome issues of lunar exploration.

NASA lost the original magnetic tapes which recorded the DDE data, but O'Brien had copies and had already started to analyse the data using Australia's first supercomputer, the SILLIAC (Sydney version of the Illinois Automatic Computer). In 1970 he published the first scientific paper to analyse dust effects from the first rocket launch from the Moon. He went back to the analysis in 2006, and in 2013 assisted by research student Monique Hollick.

The DDE data showed that rocket exhaust from the take-off of the Eagle ascent module sprayed the solar panels, and other experiments and remains on the surface, with a significant amount of dust. This caused the PSE to overheat and fail. When O'Brien and Hollick – who's now better known as an ace Australian Rules football player – analysed how the amount of dust varied over time (the DDE continued collecting data until 1977), they concluded that sunlight charged the dust and made it stickier. As the terminator (the dividing line between night and day) moves over the lunar landscape, little dust clouds filled with sharp abrasive particles dance until the charge dissipates. This has massive implications for how humans might build infrastructure for a settlement or mining activities on the Moon. It also has implications for the survival of the lunar heritage sites abraded by these miniature dust storms.

The dust detector itself could be considered a piece of Australian heritage on the Moon. It's about the tiniest of things – a matchbox experiment, a filing error, microscopic dust particles – but we'd know much less about dust, shadows and light on the Moon without it.

1970s: THE BACKYARD SATELLITE

Right in the middle of the Cold War, with the fear of nuclear holocaust ever present in people's nightmares, a network of regular people was demonstrating that space wasn't all about the superpowers. This movement was spurred by ham radio amateurs who loved to listen to the space-age sounds of the early satellites in their sheds and backyards. This is the story of a backyard satellite.

Here is what it looks like: an aluminium box weighing 17.7 kilograms, approximately 30 × 45 × 15 centimetres in size. It's painted in black and white stripes like a Collingwood football jumper. Inside, there's a battery to provide power and a transponder to send signals. The antennas are a bit unusual: they're made from spring steel carpenters' tape, just the kind you might use around your house, or as an archaeologist to measure things in the field. The tapes were wrapped around the satellite to extend on ejection into orbit. The satellite's name is Australis Oscar 5 and it was the first amateur satellite to be remote-controlled and to have a complete telemetry system. It's been circling Earth since 1970, nearly fifty years. Not bad for a satellite built by a bunch of students and their mates!

In 1958, a group of US ham radio enthusiasts formed Project OSCAR (Orbiting Satellites Carrying Amateur Radio). They launched OSCAR I, their first satellite, in 1961, only four years after Sputnik, followed by OSCAR II in 1962, and OSCAR III and IV in 1965. The volunteers weren't just building the satellites: they relied on people around the world to send data about the satellites' locations. More than 570 people from twenty-eight countries participated in tracking OSCAR I.

In the 1960s the Melbourne University Astronautical Society built their own tracking station on the roof of the Old Physics Building using old car parts, second-hand electric motors, and pieces of an air conditioning plant. As a student I walked past this building most days, unaware of its history. The student space scientists tracked Project OSCAR satellites, as well as weather satellites and US and Soviet lunar probes. But some of them were more ambitious, and decided to make their own satellite – perhaps Australia's first. The team were Richard Tonkin, Peter Hammer, Owen Mace, Stephen Howard and Paul Dunn, with help from Les Jenkins, a senior electronics technician with the CSIRO. Their supporters included the Wireless Institute of Australia and the University of Melbourne Union, who provided funding; local electrical companies who donated equipment; and various Melbourne University clubs and societies. It was a real community effort.

By 1967 the satellite was finished. At this stage it looked as if Australis Oscar 5 might beat the official satellite, WRESAT 1, to become the first Australian-designed object in orbit. But circumstances intervened and the Californian launch was repeatedly delayed, allowing WRESAT 1 to achieve that goal when it was launched from Woomera on 29 November that year. It's hard to know for sure, but there are faint hints that the Australian government didn't want to be beaten by amateurs, and may have had a hand in the delayed launch of Australis Oscar 5 …

By 1969 the Project Australis team had practically lost hope when the original Project OSCAR group merged with another to form AMSAT (the Radio Amateur Satellite Corporation). AMSAT were determined to make the launch of Australis Oscar 5 happen. Negotiations re-opened, and finally, in 1970, with the assistance of NASA engineer Jan King, Australis Oscar 5 took its place on a Thor-Delta rocket at Vandenberg Air Force Base and blasted off. Radio announcements and AMSAT newsletters alerted the global ham radio community about where to look for Australis Oscar 5 and how to send their data in. Volunteers from twenty-seven countries responded.

Even though it's still circling Earth, Australis Oscar 5 is embedded in the suburbs in places such as 'a particular refrigerator and a famous Carlton oven'. Adjacent to the University of Melbourne, Carlton was the traditional location of affordable student digs before it became gentrified. One evening's roast dinner was stymied by satellite parts being cooked at high temperature in the kitchen oven. The transmitters were tested in a backyard shared with residents sunbathing on banana lounges. The team listened to the launch via a radio link in the backyard shed of a house in the Melbourne suburb of Highett. In the Canberra suburb of Torrens, scientists supporting the project built another tracking station. It took them six months to construct an automatically controlled antenna with 15-metre-high aerials. Much of the tracking station was made from scavenged junk, including the rear axle of a 10-ton truck, and aircraft propeller pitch motors.

This 'bargain basement' approach was pretty typical for OSCAR satellites. They were frequently constructed in people's homes using everyday equipment and materials, as well as donated leftovers from aerospace industry, and small components from hardware and electronic stores: space technology on a domestic scale.

Australis Oscar 5 proved that innovation could be achieved outside established aerospace industry. AMSAT satellites were among the first to use voice transponders and pioneered microsatellite technology, which is now becoming the backbone of the commercial space industry. The amateur satellites demonstrate the diversity of space artefacts and the way everyday people can be involved in high technology. Their design and construction show what you can do on low budgets, the ingenuity of scarcity in action. When it is possible to study orbital hardware from space, these home-made satellites will stand out from the thousands of slick commercial, military and scientific satellites by their appearance. Project OSCAR initiated a tradition of volunteer, international space co-operation that continues to this day.

1980s: AIMING FOR THE PLANET OF LOVE

When I was still in Queensland madly trying to catch up on decades of space exploration, one of my first moves was to see if anyone had been to my favourite planet, Venus. While large chunks of the world were watching the race to the Moon, the USSR had another target in mind: they were aiming for Venus. Throughout the 1960s, 70s and 80s, missions were sent to fly by, orbit or land on the surface. It was highly risky, and many of the spacecraft failed. But many also succeeded, leaving a series of lonely robots as sentinels to watch over the planet of love.

Each mission failure had allowed subsequent spacecraft to be designed better and tougher to withstand the high surface temperature and pressure. The USSR applied lessons learnt from re-entry studies on intercontinental ballistic missile warheads. Design improvements included changing the core body to titanium, which has a melting point two and a half times greater than that of aluminium; more robust heat shields to protect the landers as they descended through the thick sulphuric acid clouds; and a circular ring shock-absorbing landing apparatus. This iterative design process culminated in Venera 14, the last Venus landing mission, in 1982.

The Venera 14 lander was a hermetically sealed titanium pressure sphere, which contained most of the instrumentation and electronics, mounted on the ring-shaped shock absorber and topped by a cylindrical antenna. A flat circular skirt sat around the sphere below the antenna, as an aerobraking device to slow the lander down. It also had a parachute. The shock absorber had a colour photographic scale around the edge so the photographs it took could be calibrated. It weighed 760 kilograms. To me, the successful Venera spacecraft look a little like Daleks, the *Doctor Who* cyborg villains. Kerrie Dougherty, the former curator of space at the Sydney Powerhouse Museum, explained to me that Soviet space technology didn't shield anything that didn't need it, unlike that of the US, so you can see the patterned surfaces and internal structures. It's hard to get data on how much a Dalek weighs but by one calculation the Venera is equivalent to six standard Daleks.

Venera 14's camera windows were quartz glass, made by melting pure quartz crystals to create a highly durable and temperature-resistant glass. In the images the spacecraft sent back, we see a field of flat rocks, with the curve of the shock absorber visible on the lower edge. The perspective, as if a person is looking down on their own feet, gives the photographs a personal feeling, like a selfie from space. There are few Venus analogue landscapes on Earth but the view reminds me of the gibber plains around the Woomera launch site in South Australia, with their flat pavement of brownish cobbles.

Although Venus is a close neighbour and the subject of speculation and study since ancient times, very little was known about it until the early 1960s due to the impenetrable cloud layers. Exploring Venus was considered to be critical in understanding the evolution of Earth. Our 'twin sister' has similar mass, gravity and volume. Before the first missions, Venus also held the promise of life.

Speculations ranged from imagining it as a warm, swampy world that resembled Palaeozoic Earth; dry dusty mountains; oceans of carbonic acid; and even a surface covered in hot oil or puddles of molten metals. CS Lewis (1943) created a lyrical sensorium of fragrant floating islands, a new Eden; Isaac Asimov (1954) imagined telepathic frogs swimming in Venus's warm oceans. But when the first missions returned data, the dream of Venusian life was dashed. In 1962, the US sent the Mariner 2 space probe to fly by Venus and discovered that the surface temperature was likely around 430° Celsius.

The data returned by the Venera and other missions revealed a fierce environment with the most corrosive upper atmosphere in the solar system: Venus's yellow clouds are concentrated sulphuric acid. On the surface, pressure from the predominantly CO_2 atmosphere is ninety times that on Earth; Veneras 3–6 were crushed as they descended through the atmosphere. The surface temperature is above the melting points of lead, tin and zinc. In such conditions, is it possible that the landers and probes have survived?

There is no evidence of plate tectonics and only 'modest' evidence of geological activity on Venus. Erosion processes are slow, as there is no water, and surface winds move at human walking pace. There is also little danger from the upper atmosphere. Droplets of sulphuric acid leak downwards, but evaporate as the temperature rises towards the surface – they do not survive below about 25 kilometres. Being on the surface would be like immersion in a hot dry ocean with slow currents of air. There is no reason why archaeologists of the future should not find Venera 14 exactly where it landed.

The Venera spacecraft were part of the Cold War battle to imprint space with ideology. One historian even conceptualised the Venera landers as 'Red flags on Venus', and each Venera mission carried Soviet emblems to commemorate the landing. There's only been one US surface mission, and I kind of like it that Venus still gets to be communist, just as it seems the Moon is about to be swallowed up by rampant late capitalism.

1990s: IF VERSACE WERE TO DESIGN A SATELLITE

The Cold War ended in 1991, after the fall of the Berlin Wall and the dissolution of the Soviet Union. The year after, one of the most beautiful satellites ever made was launched. Picture this: a 60-centimetre metal sphere covered in 426 diamonds. Well, they're not diamonds, but reflectors made out of fused silica, much like the Venera 14 cameras, as well as four from germanium. Each reflector is a cube corner 3.8 centimetres in diameter. The interior of the satellite is a solid brass cylinder, covered in a thick aluminium shell. It's very dense and heavy, weighing 405 kilograms. But scale it down to a centimetre and you'd be forgiven for thinking it was designed as a pendant by Fabergé or Versace.

In fact, this beauty is one of twins: there are two of these jewelled spheres orbiting Earth. The space bling twins are the LAGEOS satellites (LAGEOS stands for LAser GEO-dynamic Satellite). LAGEOS 1 was launched by the US in 1976, and LAGEOS 2, made by the Italian Space Agency, was launched in 1992. The two satellites travel at around 6000 kilometres from Earth in an almost perfectly circular polar orbit. They have no instruments or power sources; their mission is to be as stable and passive as possible so that lasers can be bounced off the diamond eyes. They don't talk or listen. Their orbits may be stable for about 8.4 million years, according to the original prediction.

Every day, thirty-five satellite laser ranging stations across the world send laser pulses up to intercept the LAGEOS satellites. The length of time taken for the two-way round trip indicates how far away the satellite is. Once the time is recorded and corrected, we know the distance to the satellite at that moment to centimetre accuracy. The changes in this distance over time relate to variations in Earth's gravitational field and rotation, as well as environmental factors in orbital space.

The information provided by LAGEOS 1 and 2 has contributed to new perspectives of Earth, as former project scientist David E Smith explains:

> Today, we see Earth as one system, with the planet's shape, rotation, atmosphere, gravitational field and the motions of the continents all connected. We take it for granted now, but LAGEOS helped us arrive at that view.

We tend to think of Earth as a perfect sphere, but the distribution of mass within it is actually rather lumpy, which means gravitational force is not equally distributed. Variations in the satellites' positions have helped scientists to accurately map this distribution and thus define the centre of Earth's mass. This is the datum for the International Terrestrial Reference System used in navigation.

Another purpose is to measure the speed and direction of tectonic plate movement, which causes continental drift. LAGEOS 1 contains one of cosmologist Carl Sagan's iconic messages to the future, wrapped around the brass core. It shows maps of continental drift, from 225 million years ago to 8.4 million years into the future, when Australia will have collided with the island archipelagos of south-east Asia. Sadly, LAGEOS 2 did not get a similar artwork. Could the satellites survive that long? The laser-reflecting diamond eyes might eventually cloud and crack, but with such robust materials, the body could survive in the cold and dry of space for a very long time, perhaps even outlasting human tenure on Earth.

2000s: A TALE OF TWO ROSETTA STONES

If LAGEOS is taking us into the future of continental drift, the European Space Agency's Rosetta mission was about looking back into the past of the solar system, 4.6 billion years ago, to an epoch before the planets existed and the Sun was circled by a swarm of asteroids and comets. One of these was Comet 67P Churyumov-Gerasimenko.

In 2004, the Rosetta spacecraft, carrying a landing craft named Philae, was launched on a decade-long chase to rendezvous with the comet. Unlike many US deep space missions, Rosetta was not powered by a Radioisotope Thermoelectric Generator (RTG); it had to rely on solar power, as did the lander. It looked a little like an insect from *Alice's Adventures Through the Looking Glass*: a square body, with a circular antenna dish peeking up like a head between two long solar panel wings. Philae had a solar-panelled body with clawed feet to hold it fast to the comet surface in the low gravity. The blue solar panels became the thematic colour of the plush toys the European Space Agency (ESA) made to market the mission.

As the intended decoder of the comet's secrets, Rosetta was named after one of the most renowned archaeological artefacts of all time, the Rosetta Stone, which provided the key to deciphering ancient Egyptian hieroglyphs. In keeping with this theme, a 7.5-centimetre nickel disc was etched at micron scale with pages of text from over 1500 languages and attached to the spacecraft under a thermal blanket. This was the result of an independent project started in 1998 by the Long Now Foundation, a consortium of linguists and native speakers working to preserve the world's disappearing languages. The technology, however, was designed at Los Alamos as a permanent data storage platform that could survive a nuclear war, with an estimated longevity of 10 000 years.

Rosetta's marathon ended when the spacecraft was successfully woken from two and a half years of hibernation as it was coasting past Jupiter. Waking up was not a given – the satellite systems could have suffered damage during their long hibernation, so the situation was quite tense! ESA started a competition where people sent in videos telling Rosetta to wake up, as if it were a sleepy high schooler struggling to get out of bed, or Jeff from the Wiggles. I was part of one of these videos: at the time I was teaching at the International Space University's Adelaide summer school, and the fabulous Italian astronaut Paolo Nespoli was visiting. Paolo gathered all the students and staff together and we yelled 'Wake up Rosetta!' at the camera. And it did.

The spacecraft went into orbit around the rubber duck–shaped comet in 2014. The next step was to release the Philae landing craft with its instruments designed to sample the surface and the coma (the cloud of gas molecules released by heat as a comet nears the sun) and surface. This was the really difficult part, as the comet's gravity would not be sufficient for the lander to stick without assistance. Along with thrusters to push it downwards, the lander's feet incorporated harpoons to drill into the surface and prevent it bouncing back into space again.

In November 2014, Philae was released, and this was where things started to go wrong with the mission. First, the main thruster failed to fire, and when the harpoons also failed to deploy, the lander did what everyone feared. It bounced – once, twice – and then a cliff stopped it in its path. It was now firmly on the surface, but the shadow of the cliff prevented the solar panels from gathering the energy needed to obtain and communicate the results of its full range of experiments to the waiting Rosetta. Nevertheless, Philae did manage to send some data back with its limited power.

What we learnt from the mission was that the comet contained a large amount of water ice – although of a different composition to Earth's water – and several organic compounds that might be called 'prebiotic'. This was very exciting stuff. We also learnt that the comet was far from a dead lump of dusty ice, with geologically active surfaces and pits.

Rosetta stayed in orbit, but Philae fell silent after July 2015. A year later, with the comet now 800 million kilometres from the Sun and the spacecraft losing power, mission control at the European Space Operations Centre in Darmstadt, Germany, started to shut down superfluous systems. This included the means to communicate with Philae. On 30 September 2016, Rosetta was sent to perform its final experiment by crash landing on the comet. It relayed images of its descent until it was 20 metres from the surface. People around the world watched the descent on live feeds and mourned the death of the spacecraft. I was not so emotional; my farewell had occurred when Philae communications were turned off. To me it seemed unbearably sad to lose the voice of a spacecraft that had been through so much.

No sensors saw Rosetta's final crash landing, but we know where it is, in a pit on the smaller lobe of the comet (the 'head' of the rubber duck). The pit is named Deir el-Medina after the ancient Egyptian town where workers on the tombs of the Valley of the Kings lived between approximately 1550 and 1080 BCE.

The estimated speed of collision was 90 centimetres per second – more or less walking pace – so it's likely that the central body of the craft is largely intact. The solar panel arrays, extending 14 metres on each side of the body, have probably snapped off and fallen into individual sheets. Perhaps some of these bounced around the pit or even outside it. Their impact will have kicked up large amounts of dust which may have settled on the exposed spacecraft surfaces.

While still in orbit, Rosetta observed eruptions of gaseous space dust from Deir el-Medina. How the materials of the spacecraft will be affected by, or will interact with, this activity is unknown. The molecules of both spacecraft materials may gradually erode, and join the exodus of dust in the comet's tail.

Comet 67P Churyumov-Gerasimenko began its life out beyond Pluto before the planets had formed. In the past it was much larger; with each pass round the sun, it loses material. Eventually it will break up and disintegrate. Already a crack is forming in the 'neck' of the rubber duck; this means that one day Rosetta and Philae may be parted and continue their journeys on separate comets. Perhaps 10 000 years from now only the Rosetta disc will remain, speaking in silent tongues, if there is anyone to find it.

2010s: THE STARMAN COMETH

Spacecraft are far more than just technology; they are woven into systems of politics, belief and emotion, and perhaps few illustrate this more than a controversial object which, I think, marks the shift from government-sponsored space exploration to private corporations and commercial interests, a key difference between space in the 20th and 21st centuries.

On 6 February 2018, Elon Musk's private space company SpaceX launched the much-vaunted Falcon Heavy rocket from Kennedy Space Center – from the same launch pad as Apollo 11 in 1969. It was a test launch which could have blown up or failed to reach orbit. Such launches generally include a dummy payload, so that the weight of the rocket is equivalent to a real launch, and they're often blocks of concrete. Musk, however, decided to have some fun.

The dummy payload he chose was his own personal midnight cherry Roadster, a sports car made by his Tesla company. A mannequin in a white SpaceX spacesuit, christened the Starman, sat in the driver's seat, its arm casually resting on the door, its face inscrutable behind the visor. For the ultimate road trip soundtrack, the car was playing David Bowie's 'Space Oddity'. There were many other symbols too: a *Hitchhiker's Guide to the Galaxy* 'Don't Panic' sign; the inscribed names of the 6000 people who worked for SpaceX; a CD with Isaac Asimov's Foundation novels in the glove box. It was a grab bag of popular science fiction references, and people loved it.

The rocket didn't explode, and the car was finally delivered into an elliptical solar orbit, its furthest point from the Sun somewhere beyond Mars in the asteroid belt. Musk thought of it as future space archaeology. As he said in a tweet:

I love the thought of a car drifting apparently endlessly
through space and perhaps being discovered by an
alien race millions of years in the future.

The Tesla Roadster might have been an expendable dummy
payload, but its primary purpose was symbolic communication. 'A red car for a red planet,' Musk had said, referring
to his ambitions for Martian settlement. But there was a lot
more going on than just the colour. There was an element
of demonstrating excessive wealth by wasting it. Giving up
such an expensive car (a new model costs US$200000) could
be seen as a sacrifice for space, but it's also like burning $100
notes to show how little they mean. It's conspicuous consumption at solar system scale, flaunting power and wealth.

In a similar vein, the red sports car symbolises masculinity – such cars are predominantly marketed to men –
but also how fragile that masculinity is. Stereotypically, the
red sports car is the accessory of choice in the male mid-life
crisis, in which men rebel against perceived domestication.
In the cultural script, the next step is often to leave their
wives and children for a much younger woman. A related
cultural meme holds that owning a sports car is over-
compensation related to performance anxiety. You know
what I mean. I couldn't help but wonder what the process
of selection had been: was there a committee of some sort?
If so, had they pointed out to Musk that these were some
of the cultural meanings of the red sports car? Perhaps he
ignored them, or didn't care. Or perhaps he was unaware
of this symbolism that circulates among women; certainly

some of my male space colleagues were very upset when I alluded to this.

In the 1960s, anthropologist Victor Turner argued that symbols can encompass two contradictory meanings at the same time. Thus, the sports car in orbit symbolises both life and death. Through the body of the car, Musk is immortalised in the vacuum of space. The car is also an armour against dying, a talisman that quells a profound fear of mortality. The spacesuit is also about death. It's the essence of the uncanny, a concept that Freud explored: the human simulacrum, something familiar that causes uneasiness, or even a sense of horror. The Starman was never alive, but now he's haunting space. He's the double or doppelganger – a harbinger of death.

But perhaps not for long. This was not a vehicle prepared for space: it was not shielded, or radiation-hardened, or made of materials that can withstand temperature extremes and constant bombardment by particles. It's likely that within a few years, all the carbon and other organic materials will have deteriorated. The interior aluminium skeleton might survive and so perhaps might a scrap or two of the red exterior. A ghost of itself, drifting in a Sargasso Sea of asteroids.

THE PHASES OF THE SPACE AGE

Archaeologists like to divide human culture into phases or ages, characterised by different technologies, industries and raw materials. Not unproblematic, as these suggest an inevitable progression of technology; all the same it is instructive to see what naming the ages of space can tell us. One schema I like was proposed by space power theorists Peter Hays and Charles D Lutes, who distinguish the ages of space by the commodities they purvey. The first period, driven by Cold War politics from 1957 to 1991, had prestige as its primary commodity. This fits with the Vanguard, Apollo and Venera missions, striving to accomplish 'firsts' in the Space Race. But little Australis Oscar 5 runs counter to this, showing that space was also about enthusiasts, makers and dreamers, who didn't have to be wealthy to participate in space.

The second Space Age, from 1991 until the present, has information as its primary commodity. One aspect of this is the information from Earth observation and telecommunication satellites that we use every day; another is the information gained about Earth and its place in the solar system, from missions like LAGEOS and Rosetta.

The third stage has wealth as the primary commodity. It seems to me that we are on the cusp of this right now, and nothing represents this phase better than the red sports car and all it symbolises.

Hays and Lutes also suggest that the next phase of space activity will involve moving away from our current geocentrism to a whole-of-solar-system perspective. The way I see it, there's a number of ways this could go. There's the idea that we're going to become a 'multi-planet species' with settlements on the Moon and Mars, maybe even further afield. Humans will adapt to different gravities, with all the attendant effects on the human body, and will have to create new architectural and cultural traditions based around the particular conditions of the planet. Some maintain that this is the only way of future-proofing the genus *Homo* from catastrophes, whether natural or of our own making.

Developing the technical capabilities needed to exploit resources available outside Earth, such as minerals on the Moon and in the asteroids, is already under way. This will radically transform the way terrestrial economies function. Soon people may think nothing of using, for example, asteroid-sourced iron in terrestrial industries. This is no longer just globalisation, but a process of interplanetisation.

Many argue that we're currently in a new era, called the Anthropocene, where geological layers will be distinguished by the marks of human activities such as industrial waste, plastics, radioactive spikes from nuclear weapons and power plants, and changes in the distribution signatures of carbon, nitrogen and oxygen. The Anthropocene has already moved off Earth and into space, with human traces on multiple planets and moons. We're changing the very nature of the solar system in the process of interacting with it through the bodies of spacecraft, sent to be our robot avatars but becoming a new type of archaeology in the process.

CHAPTER 3

SPACE ARCHAEOLOGY BEGINS ON EARTH

This morning this planet is covered by winds and blue.
This morning this planet glows with dustless perfect light,
enough that I can see one million sharp leaves
from where I stand. I walk on this planet, its hard-packed

dirt and prickling grass, and I don't fall off. I come down
soft if I choose, hard if I choose. I never float away.
Sometimes I want to be weightless on this planet, and so
…

Catherine Pierce, 'Planet'

Archaeology is a dirt discipline. In order to collect data, we go 'into the field' and walk around for days on end. We scan the ground or dig holes in it and then sieve the soil, sometimes using water to help flush it through the mesh and leave the artefacts behind. As a field archaeologist, I had a special talent: I could guarantee that on any site, at the end of the day I would be the filthiest person there, covered from head to toe in dirt and mud. I'd scrub my fingernails, and somehow fail to remove the dark half-moon of dirt that tarnished them, while looking enviously at everyone else's pristine fingertips.

Archaeology is about humans, and its scale is very human. Archaeologist Matt Edgeworth describes the physicality of excavation: in the trench we're using our bodies to excavate buildings of human dimensions, doorways that we could also walk through, and objects made to fit into the human hand, which we free from the dirt and hold in our own hands. It's very tactile. This is one of the thrills of archaeology, the feeling that you are the first person for maybe thousands of years to touch an object. Our bodies have not changed that much over the millennia, though as genetic engineering, AIs and cyborgs open the potential to remake the human entity in some other image, perhaps that's something we shouldn't take for granted.

A paradox of space archaeology is that humans barely yet live outside Earth, with only a handful of people orbiting in the International Space Station. How do you study things you can't visit, or touch? But at present, every human object in space started its life on the surface of Earth. Human material culture doesn't have to be in space to be part of the complex web of symbols and meanings that joins terrestrial culture to the off-Earth environment. Every rocket launch site says something about the values and aspirations of the nation which constructed it. A satellite in space is just a lump of matter without the tracking antennas and cameras on Earth which listen and follow it. These are new kinds of archaeological sites, not confined to one location but spread between different kinds of gravity.

If you ask a space professional why space is important, their answer is likely to refer to the fact that so many activities of everyday life rely on satellite data, from Earth observation to positioning, navigation and timing. We take it for granted that we can navigate the streets of any city in any country by using a smartphone; plan the day's activities based on weather predictions from satellite imagery; use an automatic teller machine which relies on satellite time signals. From Nairobi to Nhulunbuy, the world has become used to this kind of life.

It's probably less common to get an answer which considers the non-functional ways everyday life is entangled with space exploration. Popular culture is also part of the web of connections between Earth and space. This, as we'll see, includes food and drink, children's play and common household objects. They're all part of the world views shaped by space technology. These are also things that the space archaeologist can investigate.

When Sputnik 1 had its 50th anniversary back in 2007, my archaeological colleague and cake guru Dr Lynley Wallis made a special batch of cupcakes decorated with icing spacecraft to celebrate. I was delighted. Rocket cakes are a staple of children's birthday parties and represent youthful dreams of being an astronaut – the feeling that anything is possible, our potential is infinite, and we can reach for the stars.

The rocket cake is a charming example of the influence

space exploration has had on terrestrial food. This influence, it seems to me, falls into roughly three categories. There's food devised to commemorate space events like the launch of Sputnik 1 on 4 October, which heralds the start of World Space Week every year. Then there's food decorated or shaped like spacecraft or planets – like the rocket birthday cakes. And some food designed for consumption by astronauts in space also has terrestrial versions, like freeze-dried ice cream or Space Food Sticks. (These are still made in Australia, but the helmeted figure on the front is a BMX cyclist, not an astronaut.)

A great example of the first two is the food that arose around the launch of Sputnik 1 and 2 in 1957. These recipes and dishes can be regarded as a sort of performance, as they exist only in the moment they're put together, and disappear in the act of consumption. The practice of making space food is part of the intangible heritage of the Space Age.

THE COLD WAR STAYED FOR DINNER

The USSR's successful launch of Sputnik 1 caught the US by surprise. Suddenly, the night sky was transformed from a serene celestial dome to a place of menace, from which unseen attacks could be launched on the capitalist world. At the same time, there was tremendous excitement that the shackles of gravity had been broken at last. Human dreams of space were about to be realised.

While the US military and government were grappling

with the political implications of Sputnik 1, one of the ways in which ordinary people responded was to translate the body of the spacecraft into something familiar and edible. The humble olive, with the addition of three or four toothpicks to represent antennas, became a symbol of the satellite. This was an excellent garnish for a martini, sandwich or the quintessential American food, the hamburger.

In 1957 a newspaper photograph showed an American woman tucking into an unusual hamburger. The caption read:

> Not to be outdone – Harriet Phydros samples a Sputnikburger which an Atlanta café rushed onto the menu. It's garnished with Russian dressing and caviar, topped by satellite olive and cocktail hotdog.

Ironically, the Russian dressing mentioned is not Russian at all: it's an American salad dressing which is also used in Reuben sandwiches. The hot dog is meant to represent Laika the dog.

The 'not to be outdone' is a curious statement. Does this imply that Harriet's imminent consumption of the Sputnikburger will somehow tip the Space Race back in the US's favour? Or that eating it is a small conquest of space in its own right? In any case, this potential weapon of mass destruction has become a commodity rather than a beacon of communist ideology.

The toothpick antenna appeared in other Sputnik foods. A photograph on the cover of a 1957 *Life* magazine showed a young woman with a Sputnik sundae, surmounted by a ball of ice cream studded with toothpicks. A popular hors d'oeuvre of the 1960s featured cubes of cheese, fruit and sausage threaded on toothpicks, then stuck into oranges or melons. And of course there were Sputnik cocktails.

Sputnik cocktail

There are many versions of this about but this one seems to be the most common.

45 ml vodka
15 ml Fernet-Branca (an Italian digestive liqueur)
15 ml fresh lemon juice
½ tsp sugar

Shake all together and serve in a cocktail glass. Be warned; it is very bitter. And unless you like Fernet-Branca, the rest of the bottle may sit undrunk in the back of the liqueur cabinet for a long time.

Space historian William E Burrows, in his classic *This New Ocean: The Story of the First Space Age*, described the reaction of the US to the launch of Sputnik 1 as 'the Sputnik cocktail: vodka and sour grapes'. Food was a metaphor for the national character. At the time some US politicians, including president Dwight D Eisenhower, blamed the Soviet lead on the hedonist consumer lifestyles of Americans. As Burrows frames it, they effectively called upon US citizens 'to push away their banana splits, hot fudge sundaes, malteds, cherry-lime rickeys, barbequed steaks, and hot dogs', and make sacrifices in order to regain the upper hand in the Space Race. The Cold War came down even to what people were having for dinner.

Sputnik-inspired food, however, meant different things in different places. The journalist Fred Blumenthal, travelling though Asia following the launch, observed that 'Sputnik just turned the East upside-down. As it did everywhere, I suppose it appealed to the imagination.' In the Philippines, which had been an American colony from 1898 to 1946, Blumenthal observed restaurants, cinemas and even taxis which had been renamed Sputnik. The Sputnik toothpick olive made an appearance as a garnish on one restaurant's sandwich menu. But, he says:

> More important was the diplomatic loss of face
> suffered. In Hong Kong, Japan and the Philippines,
> there was a loss of confidence in the United States and,
> under the surface, a sort of secret glee that [it] had been
> toppled from the high horse.

So here, the consumption of Sputnik-themed food was almost an act of resistance against US imperialist ambitions. We can see in these few examples that there was an ideological dimension to Cold War food, a literal internalisation of the values that Sputnik was felt to represent for its different audiences.

The culinary legacy of the Cold War in space lives on in the 21st century. Any dish which features long spiky things can be called a sputnik; I've seen potato sputniks, a scoop of mashed potato with carrot and cucumber sticks for antennas (which sounds horribly like boarding school food to me). The famous exponent of French cooking, Anne Willan, created a baked Pineapple Sputnik with whole vanilla beans for antennas. The kohlrabi (a type of turnip) is frequently called the sputnik vegetable, as its leaves are angled just like the satellite's antennas. I'm not a fan of anything turnippy; and I regret that I don't like olives, as it's hard not to appreciate the simple elegance of the olive sputnik. Fortunately, a glacé cherry works just as well for cocktail purposes.

Is bitterness the signature flavour of the Space Age? Russian president Nikita Khrushchev disparagingly compared the Vanguard 1 satellite to a grapefruit, and I've often wondered if there was more to this than meets the eye. The grapefruit is a sort of hybrid between the older citrus varieties of pomelo and shaddock (or chadique, a much more appealing name than one which sounds like a fish), and originated in the Caribbean. One of its defining features is the bitterness resulting from a high concentration of the chemical naringin. It was introduced to Florida in the 1800s. By the early 20th century, the popularity of the fruit was growing, and the US was the main supplier to the rest of the world, having 70 per cent of the global market by the late 1960s. The grapefruit was not a popular – or readily available – fruit in Russia until recent decades.

One association people have with grapefruit is its role in losing weight. Half a grapefruit was a popular dieter's breakfast, or could be served as an appetiser with a glacé or maraschino cherry in the centre. (There's an idea for a Vanguard 1 recipe if you add toothpicks to the cherry.) The grapefruit diet, also called the Hollywood diet, emerged in the 1930s, with a resurgence in the 1970s and 1980s. Perhaps Khrushchev knew of the American trend of sticking tidbits of food on toothpicks into citrus fruit to serve with drinks. Another possibility is that he was knowingly feminising the satellite by associating it with the typically female activity of dieting, effectively calling it weak.

But perhaps he was just looking for a convenient-sized round fruit, and had none of this in mind. Sometimes a grapefruit is just a grapefruit. His comparison was an easy visual one to understand, and was quickly taken up – many newspaper reports at the time included the grapefruit without the intention of disparagement. Perhaps, for Americans, it made the satellite more relatable and accessible, as opposed to the more sinister Sputnik 1. The term lives on – I've used it myself – long after the intention of Khrushchev's remark has been forgotten. Vanguard 1 will forever be the grapefruit satellite.

A SPACE FOR CHILDREN

As well as space food and drinks, space could be consumed through the process of play. Space quickly became part of children's lives through the much-loved phenomenon of the rocket park. Rocket parks were common in the US, USSR, eastern Europe, and Australia, but seemed to be rarer in the

UK and western Europe. Perhaps the spectre of the V2 was just too close for those who had been bombarded by them in the Second World War. In Europe, certainly, post-war reconstruction continued for decades and playground development may have been considered an unnecessary luxury.

In the early 1950s, playground design took a fantastical turn with sculptural modernist structures and themed equipment, creating landscapes with sea creatures, ships and castles. Then Sputnik 1 burst into the skies and 'Sputnik-inspired' rockets joined the octopodes and submarines. One of the earliest playground rockets was designed by a sculptor and former flight engineer, John Svenson. In 1959 he was commissioned to build a 12-metre rocket by the Kiwanis Club in Ontario, California, for a playground in a local park. The cylindrical metal rocket, stabilised by guy wires, had three internal platforms caged in vertical bars to stop the children from falling out on their imaginary Moon trip. Rocket playgrounds expanded rapidly in Los Angeles and surrounding districts like Ontario, perhaps because there was an already established aerospace industry. Soon they spread to other US states.

In 1963, *Life* magazine wrote a feature article about the popularity of the new rocket parks. It noted that 160 playgrounds in the Philadelphia region were space-themed, and that children were flocking to them in record numbers. The rockets were either custom-made, like the Svenson rocket, or mass-produced by specialist manufacturers who circulated catalogues filled with space-age fantasies. The typical playground rocket, manufactured by Jamisons Manufacturing

Company in Los Angeles, was an 8-metre-high metal cage of vertical bars constructed around a central pole. Internal platforms divided the rocket into three cylindrical rooms to represent the rocket stages. These were very much passenger rockets, not for launching satellites. The platforms were accessed using a ladder through the floor. Four fins stabilised the rockets. The metal was painted in primary colours, principally red, yellow and blue, giving them a carnival feel. Frequently a slide exited from the second stage. The nose cone was pointed and solid, and the fuselage curved as it ascended, making it elliptical or (almost) fish-shaped.

These rockets were generic rather than a model of a particular one. They were symbolic and ciphered, much like 'the rocket' of science fiction writer Ray Bradbury's early stories, a featureless, interchangeable entity which has a mythic quality. Like the Sputnik olive, certain key features have been extracted and made to stand in for the whole. The curves, fins and nosecone all evoked the now-defunct V2 rocket, sliding it into a safe past where children could play without being haunted by death.

But when new playground safety standards came into force in the late 1970s, to prevent children breaking their limbs from falling onto hard concrete surfaces and tiny heads getting stuck between bars, the 1960s rockets were seen as dangerous. Many were abandoned, or dismantled, and these iconic structures began to disappear from park landscapes or started gently rusting away. As the original rockets vanished, their qualities were transferred to fairy tale castle towers. In post-1970s parks, any tower-shaped structure with a slide attached might be categorised as a rocket. Such misattribution indicates how strongly the association of rockets, parks and slides had become. In more than one instance, the name 'Rocket Park' survived the removal of the actual rocket. The absence of the artefact did not dim the memory.

THE ROCKET PARK COMES DOWN UNDER

As with so much else, the idea of rocket-themed playground equipment was imported to Australia from the US. Even though the Woomera rocket range was one of the earliest and most significant space launch sites in the world, few Australians had the chance to engage with it first-hand. It was about weapons first, and space second.

Its location in the remote interior of South Australia contributed to maintaining the high security needed for a Cold War military establishment. Most of the Australian population was concentrated along the coast so there was limited opportunity for the public to watch rocket launches from a safe distance, as they did at Cape Canaveral. Even if they were resident in South Australia closer to the Woomera range, entry to the town was strictly regulated – in case a communist spy sneaked in. Apart from reading about launches at Woomera in the newspapers, people's daily space fix came from the American sitcoms *I Dream of Jeannie* and *My Favourite Martian* – so popular they are still being aired on Australian television.

The Australian appetite for space was strong. In 1961,

Sydneysiders flocked to see model Sputniks and Vostoks at the Russian exhibit at the Sydney Trade Fair. The popularity of this exhibit led the mayor of Sydney to approach the USSR embassy to invite Yuri Gagarin, on a world tour at the time following his first orbit in space, to visit Sydney. Not to be outdone in the old Sydney–Melbourne rivalry, the mayor of Melbourne issued a similar invitation. Sadly, neither approach was successful, even though Yuri was not that far away, in Japan. Perhaps this was payback for Prime Minister Robert Menzies' refusal to congratulate the USSR on Gagarin's successful flight.

So Yuri never graced these shores, but Australians still wanted their share of space. A chance meeting between two men in the Blue Mountains of NSW opened up a new path. It might have happened something like this.

Dick West ran a metal fabrication business in Blackheath and was tired of making boat and car trailers. He wanted to try his hand at something different, and was thinking about playground equipment. One evening, Dick was having dinner at his local Rotary Club. Rotary, like the Kiwanis, was a service club, established in Chicago in 1905, and introduced into Australia in 1921. John Yeaman, the chief engineer for the Blue Mountains City Council, was also there, and they got talking. Yeaman had just returned from a trip to America, where he'd seen a rocket park. With his engineer's eye, he'd quickly drawn up a plan. Dick was enthusiastic, and together they persuaded the Rotary Club to embrace the idea of making a rocket park in Blackheath. Dick had found his new direction.

The rocket he made for Blackheath was a standard Jamison-style rocket with a slide issuing from the first level, red fins and a pointed nosecone. Installed in Blackheath Memorial Park, it was such a success that soon everyone wanted one. Dick went on to make at least thirty-eight rockets for parks across NSW, Queensland and South Australia.

This was a second rocket diaspora, where the Cold War weapon was scaled down and made palatable for public consumption. For this archaeologist, it raises identical questions to the V2 diaspora. How and why is a technology adopted, and how does it spread and change over time and space? I haven't yet embarked on the fieldwork needed to find out. I'd have to tour country towns, record the rockets in their parks, work in their archives, and talk to locals, perhaps over a quiet beverage in a pub, or afternoon tea with cake. You don't need to be an astronaut to do some amazing space archaeology.

These play rockets are as much a heritage of the Space Age as the original V2s sitting in museum collections. Playground rockets prompt strong emotions, more than traditional swings and slides. Social media posts celebrate the surviving rockets and recall the vanished rocket parks of childhood. Their appeal is multigenerational, with many stories of adults returning with their own children to play on the rocket. Just like space food these days, they're very retro-chic.

Why do people of all ages love them so much? Part of the reason is surely that the playground rockets are real and tangible. You can see and touch the rocket; it's made at a human scale, unlike the 36-storey-high Saturn V rocket which launched the Apollo missions. The rocket also plays the role of space-age fetish – a material object on to which you project your desires. Playing on the rocket allowed participation in a shared experience of space travel which was otherwise reserved for the elite – for both the children and the adults who accompanied them. This was space tourism at the neighbourhood level and at normal gravity, powered by imagination and dreams.

Now, people are bringing the rockets back. In 1997, when the Blackheath playground was rebuilt to comply with safety standards, Dick West's original rocket passed into private hands. The Blackheath Rotary Club, whose president Andrew Hancock is John Yeaman's grandson, crowd-funded a new rocket which was installed in March 2018. The rocket had returned home at last.

THE ULTIMATE ROCKET PLAYGROUND

Out in the South Australian desert, the true 'outback', the rockets weren't for play, but were just as interwoven into neighbourhood and community. In the 1960s, Woomera village had the highest birth rate in Australia. It was swarming with children, and they got to have the best rocket playground ever.

The landscape around the Woomera rocket range is punctuated with vast salt lakes, sprawling out among red sand dunes from the cold, dry period of the Pleistocene, which ended about 12 000 years ago. At the edge of the salt-water Lake Hart lie the decaying remains of two launch pads. In the 1960s, a unique hybrid rocket was launched there by a consortium of European nations called the European Launcher Development Organisation or ELDO, the precursor to the European Space Agency. Britain contributed the first stage: a silver-ridged intermediate range ballistic missile called Blue Streak. The second stage was the French Coralie. West Germany built the third stage, a rocket called Astris. This was a big deal, as Germany had been kept out of the Space Race after they had given the V2 to the rest of the world. Some said it was doomed to failure right there – humorously implying that nothing requiring such international co-operation between former Second World War enemies would work. The satellite the rocket was intended to launch was Italian.

Thousands of tons of concrete were poured to make the launch pads, which were perched on the cliffed edge of the lake among stone outcrops bearing Aboriginal engravings. The huge exhaust funnel, which gives out onto the lake, was lined with special heat-resistant tiles. The funnel was curved in a way that meant only one thing to a child's mind: a slide! The children of the Woomera township would sneak out there, probably on their bikes, and hurtle down the tiles into the crusty salt below, shrieking gleefully all the way down. Their nefarious activities escaped notice for quite a while; but soon the rocket, which was named Europa, was ready to start test launches. Journalists were invited out to view the launches and security was heightened. It was then that the children were caught out and forbidden to return.

On my first trip to Woomera, I dreamt I saw a Blue Streak rocket and searched for it in vain in the rocket park. This was the other type of rocket park: it contained the actual rockets and missiles that were tested at Woomera. The rocket that launched WRESAT 1, a donated US Redstone, lies in tangled pieces inside a wire cage. After it delivered its precious payload it was recovered from the desert and brought back to be displayed. The coat of white paint with a red kangaroo on it melted off during the launch and now you can read the big letters which say USA, revealing its origin. But no whole Blue Streaks remain in Australia; the last of them were taken to the launch site of Kourou in French Guiana, where, legend has it, they were acquired by a local farmer and turned into chicken coops. Everyone thinks they've probably rusted away by now in the humid jungle; and I'm sure that chook droppings are not very metal-friendly. But until I have the chance to investigate for myself, I'm assuming nothing.

There's still a complete Blue Streak in the National Space Centre in Leicester in the UK, rearing up to its full height, all silver space-age sleekness. Over a decade ago, part of a Blue Streak was found by a farmer in Queensland, from one of the crashed test Europa launches. The minute I saw the silver ridges in the news stories it seemed this was my dream of the Blue Streak. I'd just been looking in the wrong place.

COLD WAR IN THE DESERT HEAT

The rockets and the rocket parks were echoes of real rockets and rocket places. It comes as a surprise to many that Australia was once at the centre of the Space Age. Barely two years after the Second World War ended, Britain and Australia signed a joint agreement to develop ballistic missiles. Australia hoped, through this arrangement, to gain a greater defence capacity to fend off threats from Asia. The German V2 rocket formed the basis of this new weapon system.

The rocket range was surveyed in 1946 by the Army's Survey Corps. They started on the gibber plains around Mount Eba in South Australia. Gibber is a distinctive geological formation consisting of pebbles and cobbles lying closely together or interlocking, known as desert pavement. Over time, the action of wind and water polishes the stones and leaves a film called desert varnish. Underneath are red soil deposits, often quite deep. Gibber plains are concentrated in South Australia, particularly around Woomera. When I was doing heritage surveys with Kokatha Traditional Owners in the gibber, we'd find stone tools in great quantities lying among the pebbles. Often they were the same colour and could only be distinguished by the classic angular features of a conchoidal fracture. To my mind, the flat mosaic pavement of the dull brown gibber resembled nothing so much as the surface of Venus, as photographed by the USSR Venera landing missions.

In 1946 senior British military personnel took a flight to see the proposed area for themselves. They flew over the

Central Aborigines Reserve on the borders between South and Western Australia, the direction in which the future rockets would be launched. To their eyes, the red desert recalled the white sands around the Trinity site in New Mexico, US, where the first atomic bomb was exploded in 1944. The Australian author Ivan Southall described this view as 'one of the greatest stretches of uninhabited waste-land on earth, created by God specifically for rockets'.

There was some discussion about calling the range 'Red Sands', to echo the White Sands rocket range near the Trinity site, but in the end it was decided to call it Woomera, the Dharug word for a spear thrower or launcher. The Dharug are from NSW, but the popular Aboriginal vocabularies frequently didn't distinguish between language groups and the word was widely known across Australia. Len Beadell, known as 'the last of the explorers', surveyed the roads through the desert, where previously only Aboriginal people had known how to find their way. The range would eventually enclose an area greater than the size of the UK.

The headquarters for the Weapons Research Establishment (WRE) was a former munitions factory at Salisbury, on the northern outskirts of Adelaide. The rocket range itself extended from the rangehead area, 450 kilometres north-west of Adelaide, into Western Australia, and contained the Woomera village, nine launch areas, workshops, instrumentation buildings, hangars, tracking and meteorological stations. There were radio antennas that tracked the US Mercury and Gemini human spaceflight missions, and a Minitrack antenna for Vanguard 1.

In the 1960s and 1970s, Australia was a member of the 'Space Club', as newspapers liked to call it. Throughout this period, families were raised and gardens nurtured in the township of Woomera. Under the watchful eyes of security personnel, swimming carnivals and football matches took place; annual balls with beauty competitions such as 'Miss Guided Missile' and an endless stream of visiting dignitaries and royalty stepped off tiny planes to visit Australia's premier weapons and space facility.

HOW TO FORGET YOUR OWN SPACE AGE

Today, the huge Europa launch pads lie in ruins, riddled with holes where they were used for target practice in army training. The township population is declining and many houses and buildings are empty. Most Australians remember Woomera more for its notorious migrant detention centre than its glory days as the world's second busiest spaceport. In 1999, the Australian government built a detention centre for asylum seekers, just outside the Woomera village. Refugee families were incarcerated inside the barbed-wired fenced compound, out of sight, until 2003. The empty compound is still there and photographing it doesn't seem to be encouraged even now, as I discovered when I tried.

The desert hid Australia's Space Age from most of the population, as it also hid the nuclear testing of Maralinga and Emu Field, and later the detention centre. We see this pattern across several countries, where unpopular installations are located far from scrutiny and casual passers-by. Woomera is still a busy test and launch range, but most of the projects are high-security defence ones. You still can't wander out there and watch a launch, and even the amateur rocket societies are no longer launching at Woomera.

This is something I think about whenever I visit. It's not a closed town any more; you can go there and stay in the ELDO Hotel, which is the old quarters built for the German, French and British rocket teams that lived there during Europa launches. The hotel bar has a balcony that looks out over the desert. There is something magical about sitting there with a beer, as the sun sets over the gibber plains. It's not fancy, but I find the town's architecture, epitomising the bare functionalism of the 1960s, aesthetically pleasing. The gardens of the empty houses are neglected and dying. Water is a precious commodity here, and the predominant colours are the grey-green of hardy gum trees. The streets are quiet now and I try to imagine the hustle and bustle when 6000 people lived here, working in shifts around the clock. Once I walked down Gooyong Street, which isn't far from the Heritage Centre, thinking of Gooyong the racehorse. The residential streets are all horseshoe-shaped: this was part of the design to create a feeling of neighbourhood and community. In the centre of town, a landmark is the red nosecone of the British Black Arrow rocket which rises above the trees around the Heritage Centre.

Yet this was also a domestic front of the Cold War. The US missile tracking facility Nurrungar was just down the road from the village. It made Woomera a potential Soviet target. The antenna dishes were hidden under white radomes so no-one could guess which part of the sky they were sur-veilling. Although the facility was closed in 1999, the white domes are still visible from the road as you drive towards the Pimba Roadhouse on the way into town. It was the site of many protests during its operation. One protest took a typi-cally Aussie direction: the protesters painted a letter on each of their buttocks spelling out an anti-Nurrungar message and were photographed naked, facing away from the camera to protect their identities.

Woomera's heyday was over by the early 1970s, when the UK cancelled the Blue Streak rocket to buy US Polaris missiles, the Europa rocket moved to French Guiana, and the Apollo lunar program ceased. There was more space-age bitterness as space infrastructure was recklessly destroyed, dismantled and sold off as scrap metal by disillusioned staff. Australians had been promised a Space Age, and had kept their side of the bargain. But it seems the great 'cultural cringe' extended to space technology too. After launch-ing one Australian satellite, we retired hurt. Space became a dirty word to politicians and I've even heard people say 'Forget Woomera. It's all in the past.'

VALLEY OF THE CABLE TIES

Australia's continuing role in global space exploration after the 1970s mostly consisted of tracking the space missions of other nations. You didn't have to travel to the desert to see the tracking stations as many of them were located near populated centres. Three were near Canberra, in the Australian Capital Territory (ACT). The city also has another space claim to fame. One of Finnish architect Matti Suuronen's famous Futuro houses is located at the University of Canberra. Shaped like a UFO, it's the ultimate space-age house.

The Canberra Deep Space Communication Complex at Tidbinbilla is an important NASA facility and tracks numerous missions, including the Voyager spacecraft in interstellar space. Honeysuckle Creek tracking station, decommissioned in 1981, received the first signals from the Apollo 11 mission to the Moon in 1969. Not far away is the Orroral Valley tracking station. It operated from 1965 to 1985 and was a bit of a workhorse, focusing on Low Earth Orbit satellites. Both Honeysuckle Creek and Orroral Valley have been razed to the ground. I took a fancy to Orroral Valley, partially because it was less famous and often overlooked, and also because it was on nice flat ground – easy for geophysical and archaeological survey.

Orroral Valley was originally part of the Satellite Tracking and Data Acquisition Network (STADAN) – an international network of twenty-one tracking stations run by the Goddard Space Flight Center in the US. It had an impressive array of antennas including the SATAN (Satellite Automatic Tracking Antenna Network) antennas. The SATANs looked like a bunch of heat rays mounted on a square; they were very space-agey and suitably diabolical in appearance. Woomera's Minitrack antenna was relocated here too. Other features included the powerhouse, operations block, canteen, workshops, pumphouse, sewage treatment area, gatehouse, a lunar laser ranging facility and the Minitrack operations building. The centrepiece was a massive 26-metre dish antenna, now at the Mt Pleasant Observatory in Tasmania.

In the 1990s, what remained of the above-ground structures was demolished, leaving only the concrete footings of numerous buildings and antennas. Near the entrance to the facility, now used by tourists, hikers and other visitors, the grass is cut and the gardens sort of maintained. Deeper into the site, tall weeds and grasses are more prevalent and are invading the antenna footprints. Kangaroos have colonised the long grasses and look at you with mild annoyance as you walk around their territory, disturbing their breakfast.

It was at Orroral Valley that I conducted my first archaeological fieldwork on a space site. It started with a geophysical survey in 2009. Using a magnetometer, which can detect sub-surface metals, we found that even though the buildings had been removed, there was quite a bit left intact under the ground. This included rubbish dumps where antennas were buried, and cable trenches with cables still inside. So I started thinking about cables as space-age infrastructure, carrying electricity and data between spacecraft, antennas and computers. In 2011 I took a team of students out to Orroral Valley for an archaeological survey. This led to an obsession with cable ties which, sadly for my friends and colleagues, shows no signs of abating.

In general, surface visibility across the site is on the low side, and there's not much evidence of past human activities apart from the concrete slabs. The 'rubbish' has been cleaned up. No obvious artefact scatters; no personal objects; no bits of antenna support lying discarded. Occasionally there's a bit of recent rubbish near the picnic/parking area.

I wasn't so sure this curated surface would yield nothing to the eagle eye of the archaeologist, though. On one visit I noticed an old scrubbing brush lying outside the canteen building, the sturdy bristled kind with a wooden back, as used in a million domestic and industrial kitchens across the land. I was very keen to systematically survey the surface of the entire site to see what else we could find relating to the tracking station period, and what its spatial distribution might tell us. With the thick grass cover, I imagined it would be quickly done.

We started at what I thought would be the easiest part of the site, below the former 26-metre antenna, which is fairly thickly grassed with few weeds. I really didn't think there would be much there; I thought it would be a nice, quick demonstration of the principles of surface survey for the students.

The team walked 5 metres apart, each following a straight line or survey transect. As they walked they slowly observed the ground within their swathe of vision, and placed a pin flag at every location where they saw human material or animal scratchings/burrows. The latter was so that I could get an idea of how disturbed the surface was. We had 100 pin flags (long spikes of metal with a coloured plastic tag on top) and I thought these would last a good while. The line would move from one boundary fence to the next in formation, flagging everything of interest, and then we would look back and see how material was distributed by the density of the flags. Then, in small groups, the team would fully record each artefact or trace, including its co-ordinates, material, dimensions, shape, colour and likely function, removing the flag as each location was completed (and leaving the artefact in situ). We'd then move on to the next 50-metre traverse.

We ran out of pin flags before we had even reached the opposite fence. We could see stuff everywhere, despite the low visibility. There were bits of concrete, star pickets, bricks, lead, tin cans, insulation, wire, nails, leather, cable trenches (some of the rabbit scratches turned out not to be, once you examined them closely), metal steps, pipes, and much more. When we made a list of all the observed artefact materials or types, there were over thirty.

ARTEFACT OF THE SPACE AGE – OR RUBBISH?

Pretty soon, as the teams moved systematically through recording the artefacts, we became aware that there were quite a few cable ties (also called zip ties or ty-raps) present. So many that some people suggested that there was no need to individually photograph and measure every one. Tempting though that was, I stuck to my original methodology. There was a little grumbling. Why record the same features on so many of them, when they were all identical? What would we learn?

As it turned out, the cable ties were far from identical

and it was the minor individual variations which were the most informative. The first thing to note was that these were *used* cable ties, not new ones. They had been removed *from* something. One team observed a tie that had been torn apart, as the edges were jagged. I then instructed everyone to pay attention to the ends and record their state. Three variations then became evident: some had been torn, some cut and some melted. This tiny observation on a discarded piece of plastic translated into a decision and an action taken by a real person in carrying out a task. This was archaeology.

As we compared the cable ties we found across the erosion scar, other variations emerged. There were different colours: black, white, translucent. There were different lengths and different widths, from the very skinny and short to the very long and thick. The length and thickness could be an indication of the diameter and load that the tie had supported. Some had had the loose ends trimmed off. Some had patent numbers on them, or manufacturer's labels. Some were lying flat on the surface; others were actively eroding out of the silty soil.

This ubiquitous, seemingly simple object was raising all kinds of questions. How did the cable ties get there? Were they associated with the tracking station? What date were they from? When, exactly, had cable ties been invented? None of us knew. We all knew what they were, but we knew absolutely nothing else about this common, everyday object.

By this time someone had started substituting cable ties in film names. It was such a shame that we had no mobile phone coverage or there would have been a great Twitter hashtag in it. #Cabletiemovies such as *The Texas Cable Tie Massacre* and *The Valley of the Cable Ties* produced much hilarity. (My own effort: *The Cable Tie*, a film starring Jim Carrey.) But I could tell that despite their original scepticism, everyone was getting caught up in the cable tie story. I decided to have a brainstorm session.

What were so many cable ties doing near the main antenna? Were they associated with its dismantling? One student, Steve, imagined a bunch of blokes climbing all over the structure cutting the ties off, as they took it apart to be transported to Tasmania. The prevailing wind over the previous couple of days had been from the north-east; if this was the same in the 1980s, then perhaps the cable ties were just whisked off the ground, scattering over the grass to the south-west of the antenna. It was a plausible theory.

Tom, who had first noticed the melted ends, told us that cable ties were commonly used by hikers to secure their baggage. They would then burn them off with cigarette lighters. Just to the right of us was a path where already we had seen several groups of Outward Bounders travel down to the other end of the valley. These groups of high-school students were camping up near the old canteen, and every day two different groups would walk right past our survey area. However, as someone else pointed out, they were just walking through: you would expect to see old cable ties more at camping locations.

So we had some reasonable hypotheses here, and would need to do some research to discriminate between them. It was no good making assumptions that antenna cables = cable ties. We had also to consider that many might have washed down from further upslope, and might not have blown down from the main antenna at all. A key piece of evidence was clearly going to be the distribution of cable ties over the whole site. What features were they most associated with?

We didn't get a chance to do any more transects, but we did investigate the 9-metre antenna footing, where we found a sparse scatter of cable ties. So clearly they were present elsewhere at the site, although the association with the actual antennas was not certain since we had only looked at two of them.

Later, we visited the Canberra Deep Space Communication Complex to compare a working space tracking station with the archaeological one. Joan, another student, gave a wry smile before asking public relations manager Glen Nagle how common cable ties were in the construction and operation of the big dish antennas at Tidbinbilla.

Now, I thought, my moment of vindication. I was a little disconcerted when Glen roared with laughter! But when he recovered, he had some interesting insights. On the main 70-metre dish at Tidbinbilla, he said, there would be thousands of cable ties securing electrical and data cables. They were in fact a safety hazard – if the spiky ends are not cut off, then they could easily take out an eye. So there was another factor to consider in our assessment of the cable ties from Orroral. Then he sent his offsider to the office to get us all a genuine Tidbinbilla cable tie! Mine lived in my pencil case for a few years after that, joined from time to time by others I'd collected in the streets of various cities.

Susan was the first student to get back within mobile phone range, and she immediately texted those of us still on the road with the results of her internet search. She found out that the plastic mass-produced cable tie was invented in 1958, Year Two of the Space Age. My next mission was to find out the full story.

THE STORY OF A SPACE AGE OBJECT

Electrical lighting was introduced in New York City in the late 1890s, and an engineering company called Thomas and Betts was established to take advantage of the new market. As electricity expanded, the company began to make everything you needed to wire up something. By the early 20th century, not only buildings, but planes, ships and military vehicles had to be wired. Tape, cord and twine were used to fasten wires to each other and to parts of the structure so that the wires were organised efficiently, separated from moving parts, and traceable for servicing purposes. This worked well enough, but the string could unravel and lose tension, or cut into the wire insulation if too tight.

All of this changed in 1956 when Thomas and Betts designer Maurus C Logan paid a visit to a Boeing factory where he witnessed the wiring of an aircraft harness. Aircraft were filled with thousands of metres of wires and cables, which were organised onto sheets of plywood and secured with wax-coated nylon string. (In the 1960s, Boeing used trained ferrets to take the wires into the hard-to-access parts of the plane.) The string was knotted around the cables to hold them in place, and the tension had to be right. Pulling the knots tight cut the workers' hands, and they developed thick calluses. The result was called 'hamburger hands'; not an appealing image. Pitying the workers whose hands were being destroyed, Logan decided that there had to be a better way. For the next two years he experimented with ways to secure cables effectively, which would also be kind to the hands. He submitted the patent for the first cable tie on 24 June 1958. Thus the cable tie was launched, virtually at the same time as the first satellites. I love that this technology was invented not to solve a technical problem, but a very human one.

Of all the things I expected to get out of using a very 'traditional' archaeological technique on a space site, the discovery of cable ties was certainly not among them. My initial field season at Orroral had led me to realise the importance of cables, as opposed to the fancy stuff like antennas, but it took the application of the archaeological eye and a systematic approach to recording to tease out the implications. It's so obvious when you think about it.

One of the things that make cable ties so interesting is that they migrated from aircraft use to everyday life, and even into space. There are radiation-hardened cable ties that are used on satellites and space stations! And removing them for exterior maintenance is a problem, as those sharp severed ends could float away to join the space junk population and contribute to the risk of collisions. Many of us probably have a packet of cable ties in the garage, or kitchen drawer. People use them for virtually everything that requires fastening and much more besides: art works, spikes on bicycle helmets to ward off swooping magpies in spring, luggage ties, bird tags and handcuffs. I'm sad to say that I read *Fifty Shades of Grey*, the novel by EL James which gave BDSM a certain popular currency for a while, and watched the film, just to see the cable ties which feature in boudoir shenanigans. Now, instead of scanning the ground for stone tools, I look for cable ties. I always see discarded ones at bus stops, where they've been used to attach a poster to the shelter or a nearby power pole. Just like the Orroral Valley antennas, the poster has gone, leaving only the cable tie to betray its presence. So while I'm extremely interested in cable ties in their own right, they're also a presence which reveals an absence. I think of it as part of the science of ichnology, which studies the trace fossils of living things. Robots and computers can leave trace fossils too.

The following comes from the short story 'Open Veins' by British writer Simon Ings. It is an uncannily accurate description of abandoned space and military sites I've visited:

> The site bore little mark of its military past. The hardened bunkers, the offices and barracks, had been ripped out years ago. The radar arrays and satellite dishes had all been dismantled, leaving large, low concrete platforms, their smooth grey surfaces punctuated by rusted spars, irregular brick walls, depressions and score-marks: the tracks and spoor and burrow-mounds of artificial life. The single

concrete runway was crazed and weed-lined and
there were shreds of cable rotting in the verges.

(What's that, I hear you ask? Does he mention cable ties?
Well no, but I'm sure he was thinking about them.)

He could easily be describing the Orroral Valley tracking station here, a site characterised by the concrete footprints of long-gone satellite dishes and interferometry arrays,
which appear smooth and featureless until you start to exam-
ine them closely. Then you see the grooves left by blades
on earth-moving machinery; holes where pipes, cables and
wires vanish under the floor, weathered ledges where walls
once rose. Ings likens these to the phenomenon of the trace
fossil: the preserved remains not of some ancient creature,
but of the impressions its activities leave in the deposit that
becomes transformed into stone. They're signs, not the thing
itself. It's an appealing metaphor, to imagine the cables as
polychaete worms burrowing into the ground, the bolts left
on the antenna footings as anchors for some floating jellyfish
in its sessile phase, the hardened bunkers as coral polyps.

This flourishing fauna has come to a sudden end, ripped
out and dismantled. It's not just abandoned but actively
flattened down to ground level. This level of destruction is
frequently the fate of modern industrial sites, in contrast to
ancient ones, which are more likely to be just abandoned.
Or at least, this is argued to be one of the things that makes
the archaeology of the contemporary past different from
other kinds of archaeology. On the other hand, it's the same
taphonomy (what happens to artefacts after they have been

abandoned, through natural and cultural decay processes) that all archaeologists grapple with. While Ings says that his fictive site bears little mark of its military past, the signs should be there for those who know how to read them.

There's a complex relationship between the symbols – the rocket cakes, the playground rockets, the bitter Sputnik cocktail and tart grapefruit – and the real physical objects: the rockets and missiles, the launch pads, the antennas and cable ties. These objects exist at different time scales. Space food and drink is made over a day or so, and then consumed. The rocket is made over months and years. It's actually a very ephemeral entity, only assembled just prior to the launch, before disintegrating in the course of its launch as the stages are spent and fall away, dispersing its remains between Earth and space. While we associate rockets strongly with space, there are no whole rockets there. There are plenty of upper stages, just no rockets with all the parts present. The rocket is really a terrestrial, not a space, artefact.

The launch pads and tracking antennas are slower and heavier: they persist for decades, maybe longer, as they decay. Playground rockets are the same, but these have the advantage of staying in people's memories longer. All around them are the little minnows of the Space Age, the cable ties that are so ubiquitous we barely notice them.

There is a well-known rocket park in Deniliquin, a large NSW town about an hour's drive from where I grew up, which has an original Dick West rocket. My mother and I drove out there recently, while I was on a visit home. I watched a new generation of little kids playing on the rocket.

The slide had been taken away, and only the first and second stages were open to climb into. These kids will learn about a solar system very different from the one I knew when I was their age. For them the Moon landings are already perhaps verging on the mythical.

When Mum and I prepared to leave, my eye was caught by a familiar elongated shape in the dirt. I bent to pick it up. Yes, it was a cable tie.

CHAPTER 4

JUNKYARD EARTH

Strange flashes of radiation
zip through your ghost eyes
on this frenzied carousel
hurtling round Earth.
You wonder if radar will pick
you up as a spectral shadow
or dark mass. An unexplained
phenomenon cataloged and
monitored in the wasteland flux
where blackness leans into the soul.

Marina Lee Sable, 'Space Junk'

On Earth, space is edible, tangible and visible. You can buy
a packet of Space Food Sticks from the supermarket; hold
your phone in your hand and walk in the directions a satellite

beams down to you; take your kids to play in the rocket park. Museums and science centres often have models and simulations of spacecraft, and sometimes even unflown real ones. These are still, dead spacecraft, safe to approach and touch, just like the reconstructed skeletons of ferocious dinosaurs now frozen into meekness.

These museum spacecraft are not in their natural habitat, though, and there's no equivalent of a safari park to watch them in the wild. We can't observe them working; we only know they are there by the effects they cause, such as sending television signals. Equally, we don't see them die, decompose and become junk.

When I sat on my Queensland verandah and looked up at the sky all those years ago, I couldn't really see the space junk I knew was orbiting there. The abandoned satellites and their litters of junk were points of light with no detail against a backdrop of stars set in a black sky. Before this moment of revelation, I'd never really thought about what satellites looked like, or their individual histories. My closest encounter with a satellite at that stage was probably the community broadcasting satellite ComRadSat. Community radio stations around Australia downlinked content from the satellite, and it was the pinnacle of success to have a program picked up for national distribution. As I mentioned earlier, in Yeppoon, I presented a music program, which was called 'The World According to Alice' on Radio 4NAG, and I had a dream that I might one day reach these dizzy heights in both senses. But I didn't give much thought to the physical satellite, and in fact it wasn't even a whole one: ComRadSat was a channel hosted on one of Australia's Optus telecommunications satellites.

When GPS navigation satellites started to be used in Australia, I felt I was very brave to take a GPS unit with me into the field. The first time I did it was in the late 1990s when I was doing a survey for heritage sites on the north NSW coast. I found a very interesting place indeed: a subterranean oven, which had been operated by an Italian prisoner-of-war to make charcoal gas for vehicles during the Second World War, when petrol was in short supply. It looked like a stone-lined well and I wouldn't have known what it was if the landowner had not told me the story.

I needed to get its geographic co-ordinates so I could make a map of its location. I duly pulled out the chunky GPS unit, fired it up, and waited for it to pick up enough satellite signals for an accurate location. And waited. 'This is no good!' I cried in exasperation and no doubt thought evil thoughts about new-fangled technologies, unbecoming to one of my youthful age. I returned the unit to its box, unfolded my top-ographic map, and soon had the co-ordinates worked out. I haven't been able to stick to my Luddite ways though; now, fieldwork without accessing the GPS, Galileo or GLONASS navigation satellite constellations is unthinkable.

Even as archaeologists became more reliant on satel-lite navigation, I didn't really think about the satellites we were linking to, or about what space junk might look like. Probably most people don't: when they hear the term 'space junk', they might imagine something like a scrap yard, only floating. There were two examples that came to my mind as visual references for space junk. One was the machinery graveyard that sat outside our house paddock, where generations of tractors, trucks, ploughs, harvesters, harrowers and hay balers lay rotting through rain and shine. It was a fantastic playground even though it held some dangers in hidden snakes and sharp metal edges.

The other was a fenced area known as 'the tip', well away from the house, that we children were not supposed to visit. This was where old pots and pans, dolls' heads, broken furniture and toothbrushes ended up – I think basically any household refuse that couldn't be burnt. It was shaded by a large peppercorn tree and overgrown with thick grass and weeds. Here, snakes were even more of a danger, and the uneven layers of rubbish contained cavities that could collapse under an unwary foot. Nonetheless, we did often run to the top of the gentle sand hill and contort ourselves through the barbed-wire fence to search for treasures in the rubbish. Not far away was a large depression in the sand hill that concealed a mass grave for sheep and other animals who had died natural deaths. This was uninteresting as we were quite used to seeing sheep being slaughtered and dismembered for the weekly supply of mutton (three times a day, six days a week – I never need to eat a sheep again, frankly).

One junk pile was industrial, and the other domestic. The machinery graveyard, if suddenly blown high up into orbit and left to its own devices, might look most like what is up there now. No items of everyday life, no bones of formerly living organisms, and very little organic material. The International Space Station sends of its rubbish back to Earth, so there's no domestic refuse up there to speak of. Orbit is not much like the archaeological record of Earth at all.

One paradox of archaeology is that what is junk to most is what archaeologists find most valuable. The physical discards of someone's life become a window into their mind and into the world view of their society, revealing cultural choices that a person would not be able to articulate even if asked (as in the famous 'Cerulean Blue' scene in the 2006 film *The Devil Wears Prada,* where the formidable Miranda explains to Andi that her blue jumper was not an individual choice but determined by a broader economic and cultural system beyond her knowledge). So archaeological interests run – a little bit – counter to the environmental concerns of people who are appalled that we're trashing the solar system just like we've trashed Earth. Ancient trash is archaeology, but modern trash, the stuff we live among, is considered to have little value. But really, this distinction is arbitrary. Where does the past end and the present begin? Is it ten minutes ago, or one hundred years ago? Polluting space is no good thing, and this is among the most dangerous junk that humans have created. But this junk can also be seen as artefacts – each item the product of a distinct culture, as much so as the painted rock plaques at the Apollo 11 cave. And more than this, some of these bits of junk might have heritage value. They might be significant to living communities of people on Earth, and worth preserving into the future.

ONE THOUSAND ELEPHANTS ORBITING THE EARTH

The dawn of the Space Age was also the dawn of space rubbish in a rapidly expanding frontier of junk. Within six years of the launch of Sputnik 1 in 1957, humans had moved from Low Earth Orbit to the lofty heights of geosynchronous orbit: a few hundred kilometres to 35 786 kilometres above Earth. At this height around the equator, satellites orbit at the same rate that Earth turns: hence they always appear to be at the same location above the surface. Just three of them, poised equidistantly, can theoretically provide telecommunications coverage for the entire planet. To put this distance into perspective, the circumference of Earth at the equator is 40 000 kilometres. It's a long, long way; and it took Jules Verne's fictional traveller Phileas Fogg eighty days to go around the world in the late 1800s.

The first satellite to travel this far from Earth was the US Syncom 2 in 1963; but it fell a bit short of geostationary. This was quite an achievement as people were sceptical about a satellite reaching such a high altitude to begin with. At the Paris airshow in 1961, a prototype of Syncom was taken to the top of the Eiffel Tower. Someone quipped that this was as high as it was ever likely to get. The following year, Syncom 3 was the first to attain true geostationary orbit.

The satellite was used to broadcast the 1964 Tokyo Olympic Games to the US. This was a new era of Olympic Games coverage, which stimulated the growth of the television manufacturing industry to take advantage of satellite broadcasting capability. But it wasn't all benign. The

Vietnam War – called the American War in Vietnam – was the first one ever televised, thanks to Syncom 2 and 3, and a new type of warfare where satellite imagery helped shape public opinion was initiated. The Vietnam War is sometimes called the television war. Some of the scenes I saw on the news every night as a child were likely transmitted via the Syncoms. This was ironic, given that US president Lyndon B Johnson's recorded message, sent through Syncom 3 for the Olympic Games, spoke of how telecommunications satellites could help achieve 'lasting peace among the peoples of the world'. There was a lot riding on this cylindrical satellite, which looked a bit like a lampshade with four short legs.

Syncom 3 worked for just six years; it was turned off in 1969. By this time, it had been joined by other geostationary satellites like the Intelsat series – the corporation launched their first satellite, Early Bird, in 1965. Now there are over 500 working satellites in geostationary orbit and over a thousand dead ones.

Here's what's in orbit around Earth right now: satellites that work, satellites that don't work, the rocket stages that delivered them, bolts, canisters, fairings, exploded fragments, flecks of paint, shrapnel, tools, fuel, and, possibly, a remnant of organic waste from human spaceflight missions (yes, I mean space poo). To be classed as space debris, an object must have no purpose or function now, and none in the foreseeable future. Space junk ranges from whole spacecraft that weigh thousands of kilograms, to microscopic particles from eroded spacecraft surfaces. One calculation is that the entire bulk of this material weighs approximately 6000 tons, or by my reckoning, the equivalent of 1000 African male adult elephants. There are over 23 000 bits greater than 10 centimetres in diameter, and millions of bits smaller than that.

Orbital objects are travelling at an average speed of 27 000 kilometres per hour. If you collide with one at those speeds, there's going to be some damage. The bigger the object, the more catastrophic the effect. Each collision produces more and more bits of junk. To keep track of the growing number of bits up there, many antennas, cameras, radars and lasers on the surface of Earth are pointed up at the sky. The information about their orbits goes into vast databases. Taking into account features of the space environment such as atmospheric density and geomagnetism, the orbital data are then used to perform conjunction analyses which calculate the likelihood of collisions every day.

At lower altitudes, there is a natural cleansing method for space junk. The atmosphere, as it expands and shrinks in different space weather conditions, creates drag on objects and gradually pulls them lower. Eventually they enter the upper layers of the atmosphere at high speed and incinerate, or mostly do. Every day pieces of space junk re-enter the atmosphere and most of the time they go unnoticed. Generally, these events only receive media attention if people see it, or it's some interesting spacecraft, such as the Tiangong 1 Chinese experimental space station which fell into the ocean in 2018. The risk of being hit by falling space junk makes a good media story even though it's highly unlikely to happen. Everything in Low Earth Orbit will get dragged out of the sky eventually, but the process could take hundreds of years.

At the height of geostationary orbit, nothing is going to get pulled back to Earth. Instead, old satellites get pushed higher up and out of the way to the 'graveyard orbit' a few hundred kilometres above the geostationary ring. There's a lot of stuff up there, and we keep launching new spacecraft.

When satellites were first launched in the 1950s, people were afraid that they would drop weapons on Earth. Later the satellites themselves became the danger. As atmospheric re-entry became more common, the concern was what would happen if space junk fell out of the sky, squashing houses and people. It wasn't a completely misplaced concern. In 1978, a nuclear-powered Russian Kosmos spy satellite re-entered over northern Canada, scattering radioactive fragments and fuel across the lands of Inuit and Dene people. Operation Morning Light was mobilised to collect the toxic waste. Moon-suited workers combed the snow to find the often microscopic particles silently emitting deadly products from the radioactive decay of uranium-235. This was not a smoking chunk of space metal but an invisible contaminant at the limits of detection.

Now, the fear has taken up residence in orbit itself. The collisions we dread are no longer with Earth, but with other satellites. The danger of space junk colliding with functioning satellites and human missions is a serious problem, not just for the safe operation of the missions, but because it creates even more space junk. If we want to keep on using the telecommunication, Earth observation and navigation services we can scarcely do without these days, we need to solve this problem. There have been many proposals to clean up the mess, including space-janitor satellites, giant lasers, harpoons and nets. To date, few of the technologies tested have been successful.

The International Space Station, orbiting at about 400 kilometres above Earth, is the only permanently occupied outpost of humans in space. Inside its interconnecting tubes, a crew of three to six astronauts work. They're living the dream, but in a very smelly, noisy and dangerous environment (one astronaut said it smelt like gaol). Roughly once a year, the station has to be manoeuvred to avoid space junk drifting close to it. As well as damaging the station, a collision could pierce the hull, cause depressurisation, and risk killing the astronauts. Sometimes satellites are moved out of the way of space junk too. Billions of dollars and human lives are at stake. So far, while collisions have damaged satellites and there have been occasional catastrophic break-ups, there have been no fatalities or injuries due to space junk. The day that happens will be a wake-up call for the whole industry.

There are questions of legal liability too, as you might expect. Under the terms of the Outer Space Treaty, all objects in space remain the property of the launching state. This means that states are liable for any damage their space objects cause, in space or on Earth – if you can prove it was their junk.

My career as a heritage consultant working with engineers on mining, urban development and construction projects was a huge influence on how I started to think about the heritage values of space junk. It was no good being too theoretical and rarified: my approach had to be practical.

Doing the research to work out if space junk might be culturally significant was almost the easy part. There's no doubt that many, many satellites, like Vanguard 1, Australis Oscar 5 and Syncom 3, are bursting with significance, and this is without considering the fragments and broken bits.

In the field on Earth, there were many situations where I had to persuade an engineer (who was just trying to do their job) that the route of a power line or some other infrastructure might have to change to avoid a place of significance to the Aboriginal or European community. I had heritage legislation to back me up. There's no similar system for space.

It was very common for developers to leave the heritage survey until the last minute, because they assumed it was less important than project design. But by the time a road alignment, for example, had been chosen, there was often very little room to manoeuvre if it turned out that significant heritage was lying in its path. Changing the road would be very expensive at late stages in the design. I've had countless conversations in my career with developers grumbling at the expense of heritage assessments, as if it were the fault of the heritage. 'If you'd started this process early,' I'd say, 'it wouldn't be the big cost it is now.' Planning was the key to spending as little money as possible – which was very persuasive to developers and engineers.

The other key factor was persuading people that culturally significant space junk should not be removed from its natural setting in orbit – which would be very expensive, if necessary at all. I considered the risk of collision to see just how dangerous some of these significant old satellites might be to functioning spacecraft. Spacecraft with fuel and power can move out of the way of a piece of rogue space junk. There are also hit lists of the most dangerous junk, usually old rocket bodies abandoned in orbit which are likely to explode. It would be very useful to have a heritage list of significant satellites in orbit. If a heritage satellite appears in a conjunction analysis or space junk hit list, then we can think about how to manage it. For the moment, there's nothing we can do.

Applying these heritage principles to space junk was straightforward enough. While exploring the rich treasures of the orbital record, however, I found that I had to think more creatively about what debris, or junk, or litter was; and I also had to ask questions about how we theorised the space environment.

THE CANE TOADS OF SPACE

We talk blithely of space junk but in fact the concept of junk is a cultural one, which is not the same everywhere. We don't get annoyed at people from the past for leaving their stone tools scattered about and not cleaning up after themselves – junk is more acceptable when it's framed as archaeology. The contemporary era is different, though. It's characterised by mass manufacture and consumption, with astronomical quantities of single-use objects ending up in landfill. Materials like plastics, which don't decay quickly, or which have toxic by-products, have replaced the organics and low-level industrial products of the past. There's more stuff than there ever used to be: more coffee cups, plastic bags, chocolate wrappers, cigarette butts and paper clips.

Junk is defined not only as stuff that is abandoned, and of little value, but in contrast to nature. This is where the concept of litter adds another dimension. Litter means rubbish that has been disposed of in an inappropriate place, particularly in an urban public space but also parks, rivers, beaches and 'wildernesses' where the footprint of human activities is (usually incorrectly) thought to be absent. The more remote and less populated a place is, the more out of place a human object is considered to be – like the plastic bag found in the bottom of the deepest place in the ocean, the Mariana Trench. There are places where litter is acceptable and others where it is not. What is the proper place for space junk? You could say it is the atmosphere: that abandoned satellites and debris should be cremated, ashes to ashes, dust to dust. There's a contradiction here. We've placed junk where it is perpetually 'out of place' as a human object, but in another sense, this is its natural place.

The rise of space junk coincided with the growth of the environmental movement, which didn't view human industrial activities as the inevitable advance of human civilisation, but as destructive to Earth. These changing attitudes were partially driven by the new views of the whole Earth from the early satellites and missions to the Moon in the 1960s. People had been imagining Earth from outside for hundreds of years, from astronomer Ptolemy in the 2nd century CE to Camille Flammarion in his 1880 book *Astronomie*

Populaire. My favourite example appears in a 1965 episode of the space sitcom *I Dream of Jeannie*, when Major Tony Nelson performs a spacewalk in a silver spacesuit with a dull grey Earth in the background.

The first satellites with cameras had been sending back images of parts of Earth's surface since Explorer VI in 1959. The Soviet Molniya 4 satellite returned a picture of a grainy black-and-white whole Earth, but partially in shadow, in 1966; the US AST-3 satellite captured the complete globe in colour in 1967. In 1968, astronaut Bill Anders, on board the Apollo 8 lunar orbiting craft, took the celebrated picture of Earth rising over the lunar horizon. The most famous image is perhaps the Apollo 17 Blue Marble, from 1972, where the entire planet is lit, in daytime. For the first time we could feel Earth as a single entity – the 'Spaceship Earth'. In this metaphor, Earth is a sealed capsule of life floating in the ocean of space. The integration of humans and nature in the Earth system is defined in opposition to the menace of cold and lifeless space outside. These ideas meshed powerfully with the rise of environmental awareness.

While influential, the whole Earth and Spaceship Earth visions were not without problems. The invisibility of national boundaries gave the illusion of global solidarity and was very inspiring, but this only served to mask the fact that wars were raging on the surface, there was massive economic disparity between the 'developed world' and the 'third world', and only a privileged few – particularly the US military pilots from whom astronauts tended to be drawn – got to see this view. More particularly, they re-centred Earth, and made the Earth orbit region seem like 'outside'. They separated Earth and space and helped create a place where junk was invisible.

By 1970, hundreds of spacecraft had been launched. Some of the upper stage rockets were exploding because of the instability of their remaining fuel, creating a population of fragments. The space science community started to be concerned about the long-term impacts foreshadowed by the accumulation of this junk. In 1973, space lawyers Paul Dembling and Swadesh Kalsi talked of space debris as pollution on the last frontier.

A few years later in 1978, two space scientists predicted a frightening future if space junk was left unchecked. One of them was Burton Cour-Palais. He had worked on the problem of protecting astronauts from impacts by meteoroids, both inside their spacecraft and outside in their spacesuits. In the mid-1970s, he shared an office with Donald Kessler at NASA. Kessler had investigated the risk of collision with meteoroids for spacecraft in the asteroid belt, and he saw the immediate analogy to space junk in Earth orbit. There wasn't a lot of sympathy for focusing on a problem for which there was no solution, but Kessler wouldn't stop talking about it, earning the nickname 'Mr Orbital Debris'.

The pair argued that the real risk to astronauts and spacecraft was not meteoroids, but being hit by space junk. The higher-ups 'did not want to know about it at the time', Cour-Palais recalled in an oral history interview. So they started looking at the data about exploding rocket bodies left in orbit. They combined this with tracking data from

NORAD (North American Aerospace Defense Command) and other organisations to prove that this was a major source of orbital debris. The higher-ups had to start listening.

Then Kessler and Cour-Palais projected the data they had into the future and came up with the worst-case scenario for space junk, now called the Kessler Syndrome. A cascade of random collisions creates so much debris that Earth is cut off from space, even if nothing new is ever launched into orbit. Each collision creates new pieces of junk, which collide with each other in ever-increasing numbers. The paper, 'Collision frequency of artificial satellites: the creation of a debris belt', is elegantly and succinctly written, its strongest points delivered matter-of-factly. Kessler and Cour-Palais state that if their scenario comes to pass, 'all missions would have to expect damage in certain regions of space'. At its most extreme, any space vehicle trying to leave Earth would be smashed into smithereens. It's the robot equivalent of the Greenhouse Effect – a runaway feedback mechanism that once started can't be halted. From this perspective perhaps I shouldn't compare satellites and space junk to elephants: perhaps the more appropriate analogy is cane toads. They were introduced to Australia in 1935 to control pests in Queensland sugar cane crops, but have got out of control and are spreading everywhere. I needed help to convert the weight of space junk into the equivalent number in cane toads but here for your edification is the number: there is 8.4 million cane toads' worth of junk up there.

The idea of a cascade of collisions came from theories of planetary formation, where collisions between bits of space stuff cause the eventual formation of a ring or belt, just like we see around the ice and gas giant planets of Neptune, Uranus, Saturn and Jupiter. Eventually, Kessler and Cour-Palais proposed, as orbits decayed and became more similar, Earth might acquire visible rings, the only inner solar system planet to have them. I can't help thinking that this inadvertent work of planetary engineering could be quite beautiful.

How would it feel if the Kessler Syndrome became reality? Canadian science fiction writer Karl Schroeder explored this in a short story called 'Laika's Ghost' that describes a world where access to space has been closed off by catastrophic space debris events. One of the characters, Ambrose, laments:

> Then when I was twelve the Pakistan-Indian war
> happened and they blew up each other's satellites.
> All that debris from the explosions is going to be up
> there for centuries! You can't even get a manned
> [sic] spacecraft through that cloud, it's like shrapnel.
> Hell, they haven't even cleared low Earth orbit to
> restart the orbital tourist industry. I'll never get to
> really go there! None of us will. We're never getting
> off this sinkhole.

Earth governments – 'they' – appear helpless, unable or unwilling to initiate a clean-up operation. It's a bleak vision of what many predict will happen if nothing is done about orbital debris, and, just as importantly, if we don't prevent war in space. In 2007, China fired a missile at one of its own satellites to see if they could hit it from Earth. Well, they could; but the break-up of Fengyun 1C created so much debris that experts said we were now twenty years closer to the Kessler Syndrome than if the test had never taken place. So when Schroeder talks about war in space, we already have an idea of how that might play out.

THE COSMOS IN OUR BACKYARD

Once I had started to get a handle on what was orbiting above our heads, I realised that space junk was an example of what archaeologists and heritage managers call a cultural landscape, an ecology of artefacts within their environment. Basically, what we see when we look at the world around us is the combined effect of human activities and the natural environment. Both act together and influence each other, to form a new entity that has value in its own right. The gently undulating sand hills of the front paddocks at home were a 'natural' component of a landscape that was also incised with two deep grooves from generations of vehicle tyres which daily made their way across the property. One night as adults, after an unfortunate Christmas encounter with a bottle of Emerald Cream liqueur, my sister and I walked out along this dirt track and sang Harry Belafonte songs at the tops of our voices under the stars. The pasture was grazed by Corriedale sheep, the same kind whose pale curved ribs were piled together in the grave pit. Old fence posts from an earlier era of land management straggled across the landscape with bits of rusting barbed wire threaded through them. The past and the present were all on the same surface. Space junk was the same kind of cultural landscape, where the orbits were the grooved tracks. If I focused on individual spacecraft and their heritage values, I was missing half the picture. I had to look at space junk as a collection or assemblage, the relationships between the objects, and the interaction with the 'natural' environment of near-Earth space.

We don't think of the space environment in the same way as Earth's. One reason is the common perception of space as a black, empty vacuum. Unlike Earth, space is infinite – beyond our sun there are billions of others just like it, even in our 'unfashionable end of the western spiral arm of the Galaxy', as Douglas Adams called it in *The Hitchhiker's Guide to the Galaxy*. Perhaps most importantly, as far as we know, there is nothing living in interplanetary space that humans haven't put there. We have managed to expand the human biosphere just a little, into Low Earth Orbit, where the International Space Station circles with its tiny crew. And I hate to tell you this, but some fool sent Madagascar hissing cockroaches into space on an experimental space station that's still in orbit. There's no way that's going to end well.

Earth is slowly eroding into space with the materials that we send into orbit and beyond. We've even increased the weight of the Moon, Mars and Venus by a nanofraction. But Earth is also aggrading as far huger quantities of cosmic material fall to Earth every day – an estimated 40 000 tons each year. This interchange of material between Earth and space is a good illustration that Spaceship Earth is more like a shoreline onto which the driftwood dusts of the cosmos wash.

A wide range of extraterrestrial phenomena have a direct impact on our experience of life on Earth. These include solar cycles and solar flares, tides caused by the combined gravitational action of the Moon and the Sun, and subtle gravitational influences from other planets and the wider galaxy. The idea that Earth is an open system interacting with the broader interplanetary and interstellar environment is far from new. Many Indigenous and 'pre-modern' cosmologies accord celestial bodies a far greater role in human affairs than contemporary Western industrial societies. In these world views, space isn't necessarily 'outer'.

In the 1950s, the scholar and writer CS Lewis told his students to walk out at night and imagine how a medieval person would perceive the stars. In the modern era, Lewis said, we understand ourselves to be looking outwards. By contrast, a medieval observer would have understood gazing at the night sky to be looking inwards. This was far from an empty space: it was a living cosmos with spiritual dimensions and harmonies between the celestial and terrestrial worlds.

The separation of Earth and space was partially a result of the Scientific Revolution of the 16th century. Space became a geometry rather than the medium of life. Under the influence of the nascent sciences, the idea of space as an empty vacuum caught hold. Lewis didn't think this was a good thing. In 1938, twenty years before the launch of Sputnik 1, he described the impact of this division in *Out of the Silent Planet*. His hero Ransom is kidnapped by evil scientists (of course) and taken into space in a spherical spaceship. What he experiences surprises him:

> A nightmare, long engendered in the modern mind by the mythology that follows in the wake of science, was falling off him. He had read of 'Space': at the back of his thinking for years had lurked the dismal fancy of the black, cold, vacuity, the utter deadness, which was supposed to separate the worlds.

Instead, he experiences space as the source of life and light: blazing stars, vibrant energy, and worlds filled with life. Lewis wanted to make space come alive again and I feel much sympathy with this perspective.

When I first started thinking about space junk, I felt that an archaeology of space needed to close the gap between the 'inner' space of Earth and the 'outer' space of orbit. We needed to transform it into a place where we could under-stand human interactions with the space environment in a meaningful way. 'Nature' and 'culture' shouldn't be separ-ated; space junk was now as much a part of space as the comets and meteoroids. You could call it a non-biological ecology, or a cultural landscape, as I do.

The impression science had given us of empty space was not a very accurate way of describing this environment, as the early satellites confirmed with their measurements. More like Lewis's vision, space turned out to be a very dynamic place. Human artefacts hurtle through a matrix of cosmic particles and dust, electromagnetic currents, plasma clouds, meteors and atomic elements (on Earth it's O_2; in space it's mostly just O). Satellites, rocket bodies and other objects decay and fragment, and the materials of Earth migrate into the gases of space and create satellite dusts. Aluminium is the third most common element in the Earth's crust, but was a rarity in space. Not any more: we've put a few thousand tons of the stuff up there, and a whole range of other materials too.

Human material isn't in nice convenient chunks which we can just remove to return space to a pristine 'wilderness'. It's fractally intertwined with the very fabric of interplanetary space. If we recognise that near-Earth space is now a complex mix of human and non-human elements, the problem of space junk needs to be approached within the context of this new type of environment.

ENVIRONMENTAL MANAGEMENT IN SPACE

On Earth, my heritage work was often part of an Environmental Impact Study (EIS), whether for a road, mine, dam or residential development. An EIS considers all kinds of impacts: heritage, flora, fauna, noise, dust, water, employment, fire management, waste disposal, and many more. Heritage was one small part of working out how to make sure that the development, whatever it was, didn't do more damage than good. Governments and developers often talk about these processes as unnecessary red tape getting in the way of economic growth, but it's all about making sure that as few people or things as possible are left worse off when something is built. I'm not saying the environmental impact process is perfect, but without it – well, you can imagine. Profit would trump every time.

For industry in space, like launching a new satellite constellation or developing a lunar settlement or mine, all of this is stripped away to bare bones. The impacts considered are liability for damaging another nation's space assets when your space asset collides with it, or the impacts of rocket exhaust fumes on the upper atmosphere, which may have a long-term effect. People might reasonably object to space activities damaging the atmosphere or believe that adverse impacts on terrestrial ecologies are not an acceptable price to pay for a space industry. It's pretty clear that impacts are only considered where they might affect living things.

In orbit, the equivalent of environmental management is 'space situational awareness'. This military concept means being aware of everything around you so you can make the right strategic decisions now and into the future. It means knowing where things are in orbit – the functioning satellites and the junk, and understanding conditions, like space weather – in order to safely maximise the human use of space. Space situational awareness is the framework that people are using to try and mitigate the impacts of space junk. This is all good. However, there are no ethical obligations embedded in this concept in the same way that there are for terrestrial environments, and there's little room for heritage or social impacts.

I suppose the difference between space situational

awareness and environmental management is something like this. Think of someone running a high-precision Internet of Things operation on the farm where I grew up. An automated farm vehicle is using a sophisticated navigation device and all kinds of sensors to traverse the sand hills. It takes water measurements from the soil and determines that driving on the track after recent rains might be a bit risky and lead to getting bogged. (My father was the champion of getting bogged, and it was usually my mother who had to pull him out with the tractor.) It would pick up the location of 'the tip' and the obstacle course of the farm machinery graveyard. All these things are noted and avoided on the way from the house paddock to the back paddocks. In a remote office somewhere, a person receives the log of the vehicle's trip and thinks, 'We need to get rid of that junk.'

The first thing to note is that the presence of 'the junk' did not impede the vehicle's journey; it simply had to avoid the old rotting machinery and the little fenced-in paddock. No damage had been caused, no-one and nothing was hurt. The assumption is made that because it's abandoned, it has no value. But unlike litter, it's not out of place: it's as much in its place as it could be. The vehicle could detect the locations and materials, but it couldn't relay the history or the meanings of the objects or places. It didn't register that the machinery included a stump-jump plough from the 1870s and a 1980s Massey Ferguson header, or that little kids played on these silent behemoths in the 1970s. The junk in 'the tip' was not seen as the accumulated evidence of generations of women marrying into the land and battling isolation, mouse plagues, children dying of measles before the vaccines were invented, poverty when the rains didn't come, and working from dawn until midnight seven days a week. Beyond the limits of the vehicle's sensors were the stone tools discarded by Wiradjuri people before they were driven off or killed in frontier violence. Without a human eye to see the tell-tale angles and curves of human manufacture, they're just broken rocks, and the hearths where people cooked dinner and chatted about the day before turning in could be from any bushfire event.

If this junk was 'cleaned up', the landscape would be less rich, but no less shaped by human hands. This doesn't mean that the objects need to be conserved, though. The Burra Charter says we should do 'as much as is necessary and as little as possible' to retain cultural significance. Every family farm has its machine graveyard and rubbish tip. It's OK to let them decay, to let the wind and the rain take them.

We don't need to destroy everything currently classed as space junk either – over 95 per cent of all the stuff up there – to reduce the risks of collisions from orbital debris. We can do it in a smart way by thinking through all the heritage and environmental issues.

To study the junkyard of Earth orbit, I use documents and images that record the early spacecraft, as if they were stone tools, and use them to reconstruct the rest of the story. In other words, the world view that made a spacecraft look like *this* and not *that*. I have to think about what they meant back then, and what they mean now, as well as their significance as individual objects and as an assemblage. To work out what they look like en masse, my tools also include tracking data, simulations and visualisations.

So far, the existence of cultural heritage values for the stuff we now call space junk has barely been considered. But when you think about it, some of the artefacts that hurtle through space, far above our heads, are among the most significant in human history. Those artefacts represent the technologies and trajectories that shaped the world we live in, the era in which humanity truly became spacefaring.

Heritage management is now a routine part of any terrestrial industry or development, and space industry should be no different. This doesn't have to mean compromising mission safety or the space environment – we just need to plan for the removal of space junk in the right way. Not all space junk was created equal; some bits are clearly more significant than others. Part of what makes these satellites significant is that they are still up there. Do we really want to send them back into the atmosphere or remove them from their original orbit if we don't have to? No matter what kind of orbital debris removal scheme is implemented, it must be designed to avoid operational spacecraft anyway, so it's a small leap to ensure precious artefacts are also preserved in their natural setting.

It will be a while before we see large-scale space debris removal. We should use this time to plan a cultural heritage management strategy that will be both effective and practical. Our approach should be not to look at the *amount* of debris in orbit, but at the *risk* instead: how likely is a collision with space junk? Most of the space junk in Earth orbit is actually very tiny – millions of fragments less than a centimetre in size. Collisions with the small stuff happen constantly but don't generally cause mission failure or explosion. The middle-sized stuff (1–10 centimetres across) is more of a problem: there's a higher likelihood of collision, and the damage caused will be greater. However, whole spacecraft are numerically in the minority so while a collision would be catastrophic, it's much less likely to happen.

The risk factors also depend on the orbit, as some orbits have a much higher density of debris than others. For instance, near-circular orbits below 2000 kilometres, at around 20 000 kilometres, and at 36 000 kilometres from Earth have the highest density of large debris over 10 centimetres. The risks of collision are obviously greater in these locations.

If we're going to make decisions about what to destroy, let's do it from an informed position. We need to know which objects do have cultural significance in orbit, from local, national and global perspectives. And we need to understand how their changing orbits may relate to collision risks. But before we do this, it might be as well to think more about the definition of space junk as something that does not now, or in the foreseeable future, serve a useful purpose. Isn't representing the cultural heritage of a nation or a community, and making people feel involved in space, useful? If we acknowledge this, then many of the old defunct satellites have a very important function indeed. Like the gods who don't die as long as people believe in them, you might question whether these satellites are really as dead as they seem.

WHAT IS DEAD CAN NEVER DIE

I used to assume that satellites classed as space junk were all 'dead' and incapable of functioning. But now I'm not sure it's that simple. When, actually, is the moment of satellite death? Are they dead when they stop transmitting or when no-one is listening for their signals at the end of the official mission life? Perhaps it's when the battery or the fuel runs out. Some spacecraft have RTG (radioisotope thermoelectric generator) nuclear power sources, which can keep going for a long time; and others can go into hibernation with all the systems powered down, like the Rosetta spacecraft did for years. Are they dead when they're lost? There are many spacecraft which have slipped out of the tracking networks like shadows at sundown. Occasionally we find them again. The mysterious solar-orbiting object J002E3, identified in 2002 and originally thought to be an asteroid, turned out be a rocket stage from the 1969 launch of Apollo 12. Maybe satellites are dead when they start to break up. But how small do the fragments have to be, and which fragment contains the 'brain'? Is the ultimate death when they're forgotten – when the communities who made and used them don't care any more?

This suggests that the time at which satellites become non-functional is a continuum rather than a moment, very much like the death of the human body. This is contentious because, as it turns out, the point of human death can be quite difficult to determine. Absence of heartbeat and breath, fixed pupils and the onset of putrefaction have been the traditional signs of death. But these are by no means foolproof. As a fictional example, think *Romeo and Juliet*, where a potion made life mimic death, and mistaking the signs of death had tragic consequences.

In the 18th and 19th centuries, people were so afraid of the signs of death being misread that they rigged up elaborate systems in case they were buried or entombed alive. Safety coffins were invented with air tubes and ropes to ring bells outside to alert people to come and rescue them from a second, more terrifying death. Medical advances have increased precision in detecting brain activity, and have also determined that different parts of the body can die at different times. One can only pursue the organic death analogy so far, but flatlining, when the signal transmitted and received from the brain or heart ceases, is a bit like when a satellite transmission fails. Both are about invisible electromagnetic signals that need to be read by another machine.

This liminal space between life and death has led the popular press to coin the term 'zombie satellite' – dead but they still won't die, roaming orbit mindlessly to cause destruction. But can old satellites be resuscitated and rescued from their zombiehood?

In fact, there have been a few satellites which have been contacted or revived long after they were supposedly dead. In January 2018, amateur radio astronomer Scott Tilley found that an old NASA satellite was still transmitting data. IMAGE was launched in 2000 to map Earth's magnetosphere. Contact ceased in 2005: NASA knew where it was, but stopped listening for it. For thirteen years IMAGE orbited by itself, unmoored from its scientific team on Earth, who assumed that power had been cut to the transponder. Even better, although the overall condition of the spacecraft was unknown, it looked like the data Tilley intercepted from the satellite was still useable. How many other satellites like that are out there? A satellite that can be spoken to, listen or which has power left, isn't technically junk. The problem here, however, is that without a budget and personnel, the mission can't just be picked up again: staff and resources have been allocated to live missions. This is definitely where community scientists can play a role. Imagine if we developed a catalogue of spacecraft that could be contacted or repurposed. Such 'adaptive re-use', as it's known in the heritage world, would reduce the need to keep building and launching new spacecraft a little, and help retain the cultural value of the spacecraft already in orbit. If a spacecraft is remembered, it is definitely not yet dead. Among the paths of Earth orbit are those travelled by human memories and emotions.

'AND WARM WITH HUMAN LOVE THE CHILL OF SPACE'

Digitally generated visualisations of space junk show Earth surrounded by a thick cloud of white dots, each representing a whole satellite or fragment large enough to be tracked from Earth. This is, hopefully, a 'before' view that we'll look back on in fifty years with both wonder and disgust. The 'after' shot should look very different. Sparser. More evenly distributed, perhaps. It may include high densities at Lagrange points, places where the gravitational forces created by the Earth, Moon and Sun are equally balanced. These may be our space museums of the future, where we can park space junk safely out of the way and know that it will stay more or less put.

Some of my favourite spacecraft, such as Vanguard 1, the oldest human object in space, and the amateur satellite Australis Oscar 5, are technically space junk. The status of others is more ambiguous. LAGEOS 1 and 2, the jewelled satellites, will continue working as long as there is someone on Earth to shoot a laser at them, a more difficult future to predict. But let's meet another space object with a story to tell.

The TRAAC (Transit Research and Attitude Control) satellite was launched in 1961 by the US and is expected to remain in orbit for several hundred more years. The satellite was damaged by the Starfish Prime High Altitude nuclear test, which detonated some 600 kilometres below it in 1962, taking out its solar panels. TRAAC is part of the evidence of Cold War nuclear weapons testing, at a time before the Outer Space Treaty established space as a global commons to be used for peaceful purposes. So TRAAC represents a vision of space very different from how we conceive of it now.

The thing I really love about TRAAC is that it carried the first poem into space. The poem was written by a Dante scholar, Thomas G Bergin of Yale University. It views human spacecraft as weapons against the gods, who have until now had us at their mercy. But it assumes the best of human nature. This is how it ends:

Fear not, Immortals, we forgive your faults,
And as we come to claim our promised place
Aim only to repay the good you gave
And warm with human love the chill of space.

This last stanza was engraved on an instrument panel and became part of the fabric of the spacecraft. It was the start of a long tradition of incorporating an element of human art and symbolism into robotic spacecraft, either as messages to those beyond us in space or to those of us who remain here on Earth. The final line is compelling; it reminds me that we can choose the values we project into space, and that warmth does not come only from the Sun. If we want CS Lewis's vision of a vibrant space, then it's within our power to make that happen. In the words of astronaut Dr Amelia Brand (played by Anne Hathaway in the 2014 film *Interstellar*):

Love is the one thing that we're capable of perceiving that transcends dimensions of time and space. Maybe we should trust that, even if we can't understand it.

CHAPTER 5

SHADOWS ON THE MOON

The grey unknown
was acceptable for so long
but then we got close
and color leaked in

Christine Rueter, 'Color Leaked In'

Next time you look up in the sky and see the Moon, perhaps a full disc illuminating a dark night, or a sliver of silver suspended in pale evening blue, think about this. The Moon is a battlefield of competing ideologies: it's a strategic military base vs a romantic lovers' lamp; a scientific triumph vs government hoax; a resource to be exploited vs spiritual icon. More than anything, perhaps, the moon is magic vs science.

Nothing captured these contradictions more than the US sitcom *I Dream of Jeannie*. It was a weird kind of time slippage: I had witnessed the Apollo 11 landing with my own eyes in 1969, but in *I Dream of Jeannie* it hadn't yet taken place. Throughout the 1960s and 1970s, we watched the handsome astronaut Major Tony Nelson and his bumbling sidekick Major Roger Healey grapple with the whims of a genie straight out of *The Thousand and One Nights*, while they trained to go to the Moon. It was masculine science against feminine logic, and Jeannie won every time. If anything, I was more interested in her than in him, as I was enamoured of story-telling and the world of Scheherazade. The message was clear, though: if you tried to restrict the Moon to apparent rationality, the magical would keep bubbling through.

The prospect of space travel, as it emerged in the post-war world, gave the Moon a military function it had never had before. In *The Complete Book of Space Travel*, a 1956 book aimed at boys, Albro T Gaul said:

> Today, space travel is one of the ultimate goals of
> scientific and military research. The familiar cry,
> 'Who rules the moon controls the earth!' reflects our
> readiness to exploit space. Our military might is ready
> for space; our economic strength is ready for space;
> soon our ships will be ready for space.

By which he meant, of course, American spaceships. Albro T Gaul was an entomologist who recorded insect sounds, including 'butterflies and dragonflies in flight, flies caught on flypaper, Japanese beetles on a rose, and insects walking and chewing'. Who knew mild-mannered Albro was harbouring such bellicose thoughts? He chose to write about space because he had an unusual perspective. In the introduction to the book, he says 'space travel is also a biological problem, even perhaps to a greater extent than it is an engineering problem'.

And indeed, the role of the Moon in human affairs is intensely biological. The Moon is essential to life on Earth, controlling light, tides and time. There is no living thing which has not been influenced by it; no person on Earth, from our hominin ancestors to the present day, who has not looked up at the Moon in the night sky, experienced lunar cycles, or felt the effects of the tides. The Moon has always been a huge part of terrestrial ecology and of human culture. It's the inspiration for stories, myths and science about how the heavens and Earth came to be. And of course, speculation about how we might reach the great pearl of the night.

In earlier periods of science, we didn't know the extent of Earth's atmosphere. For all we knew, it could have reached as far as the Moon, so there was no reason a living being could not travel there. That the space in between Earth and the Moon was inhospitable to life was not obvious until the 17th century. In 1648, the French scientists Blaise Pascal and Florin Périer built a mercury-filled barometer to measure air pressure, and climbed the highest mountain they could find. As they ascended, the mercury expanded, indicating that the weight of air above them was decreasing. Pascal reasoned that there was a vacuum above the atmosphere.

The Montgolfier brothers' invention of the hot air balloon in 1783 enabled further exploration of the higher reaches of the atmosphere. Low pressure and lack of oxygen had terrible effects on the human body, but from the 1890s, when weather balloons were invented, it was possible to get data from even higher up without risking death. It became clear that Pascal was right. The atmosphere became thinner and thinner and the air pressure lower and lower, until the brain could not get enough oxygen to support consciousness. No-one would be able to breathe in space. The problem wasn't just getting to the Moon, it was staying alive during the journey.

However, the numerical distance to the Moon is not always the same as the conceptual distance. This was something that I discovered when a 'Super Moon' event occurred a few years ago. A Super Moon is when a full Moon coincides with its closest approach to Earth, resulting in a very large and luminous appearance. I received a few media enquiries wanting comments about how the full Moon affected human emotions – madness, lust, menstruation, werewolves, and so on. Certainly rich material, but I didn't want to play into misconceptions about mental health, particularly in areas outside my expertise. I decided to seek out other ways that humans had interacted with the Moon and came across some stories that surprised me.

WHEN BIRDS MIGRATED TO THE MOON

Unlike the stars, the Moon is close enough for people to think of travelling there. My favourite method of space travel was devised by French writer Cyrano de Bergerac. In his satirical work *A Voyage to the Moon*, published in 1657, he used impeccable logic. Observing that every morning as the Sun rises, dew is drawn upwards in the process of evaporation, he decided to take advantage of this:

> I planted my self in the middle of a great many Glasses full of Dew, tied fast about me; upon which the Sun so violently darted his Rays, that the Heat, which attracted them, as it does the thickest Clouds, carried me up so high, that at length I found my self above the middle Region of the Air.

Cyrano does not say how he collected enough dew for this enterprise; I imagine him alone at dawn in a field of flowers, carefully tipping each leaf and petal so the precious drops slide into the glass vessels, which he seals with a cork when full.

Once aloft, navigating wasn't as easy as he had imagined: instead of carrying him straight upwards to the Moon, he found himself moving further away from it. He was forced to break some of the glass vials to descend again. So he didn't get to the Moon by this method, but perhaps he experienced what a moonwalk might be like, as the remaining dew bottles gave him a nice bounce when he got back to Earth. I imagine this as being like the video footage of the Apollo astronauts playing in one-sixth terrestrial gravity, where they look a little like bipedal lambs gambolling just for the fun of it.

On Cyrano's next attempt, it was tiers of fireworks – a staged rocket – that propelled him to the Moon. There he met a little old Spanish man, who had arrived at the Moon in an equally fantastical manner. I did a double-take when I realised what this meant. Someone had got to the Moon before Cyrano. How, and why?

The story of the Spaniard is not just about travel to the Moon, but also about a terrestrial phenomenon that was one of the great scientific mysteries from the ancient Greeks to the 17th century: where did the birds go in winter? This was an annual event: as the winter approached in Europe, numerous species of bird flew away somewhere or just vanished. No-one knew where they went. In the 4th century BCE, the philosopher Aristotle had two theories about this. He postulated that they hibernated during the winter as other animals did. Swallows, for example, encased themselves in little balls of clay and sank out of sight to the bottom of swamps. His other idea was that the missing species transformed themselves into the birds that did stick around for the winter, and changed back when summer came.

The little old man in de Bergerac's story was an imagined Spanish soldier called Domingo Gonsales, and he was the hero of another story. In 1638, the English cleric Francis Godwin published *The Man in the Moone*, a fictional account of Gonsales' lunar adventure. In the book, Gonsales trained twenty-five swans to pull an 'engine' he had made. One day, he took a jaunt in his swan carriage which happened to coincide with the time birds were accustomed to disappear, as it seemed, from Earth.

Gonsales was about to find out the answer to the mystery. To his surprise, the swans flew upwards, until they reached what we would think of as orbit and became weightless. Pascal's work on the lack of atmosphere in space had not yet filtered through to Godwin, as both birds and man breathed as usual. In twelve days they reached the Moon – and there he found other migrating terrestrial birds, such as swallows, nightingales and woodcocks. When the swans started to show signs of agitation, he divined that they were ready to return to Earth; and so he harnessed them again and sailed home in nine days, gravitational pull on his side.

This was a ripping yarn for sure, but some thought it was a plausible alternative to Aristotle's theories, especially as there was a Biblical passage that seemed to allude to it. In the King James translation, it goes:

> Yea, the stork in the heaven knoweth her appointed times; and the turtle and the crane and the swallow observe the time of their coming (Jeremiah 8:7).

In 1684, an anonymous professor, most likely renegade physicist Charles Morton, published 'An Enquiry into the Physical and Literal Sense of that Scripture'. It seems he'd read Godwin's tale of the lunar swan vehicle. He advanced a scientific argument that birds really did winter on the Moon. And if birds could do it, then surely humans, though unwinged, could follow.

Morton rejected Aristotle's widely accepted hibernation theory, and pointed out a major flaw in the theory that the birds simply migrated to another place on Earth: no-one in Europe knew where they went. They literally disappeared. He argued that returning birds, like woodcocks, appeared to drop suddenly from the sky over ships at sea. Their round trip to the Moon took one month each way, taking the distance to the Moon and the length of their absence into account. There was no atmospheric resistance to impede their flight (so he had taken on board that much of Pascal's conclusions) and the journey between the worlds was aided by lack of gravity. They slept for much of it, living off their body fat. It was all logical enough, in its own way.

There were many factors that led to the recognition that birds migrate to other continents rather than the Moon, but one which is pretty astonishing came about in 1882 in Mecklenburg, Germany. Someone shot a white stork, and when it fell, they saw that the stork had survived a previous attempt on its life. There was a spear embedded in its neck. Intrigued, the hunters took it to the nearby University of Rostock, where the spear was identified as 'Central African'. The stork had, therefore, flown 5000 kilometres pierced with an 80-centimetre iron-tipped spear. It became known as the Pfeilstorch (arrow-stork). It was taxidermied and is in the University of Rostock's zoological collection, where you can see it today. I can't help wondering, if there were experts in Central African material culture in the university, how was it that no-one had ever observed the presence of storks there before?

The Pfeilstorch's spear was the equivalent of a bird tag. The practice of bird ringing to follow their migration routes began in 1901; and now birds are tagged and their movements tracked by satellite. Their tag? Often, it's a cable tie.

In October 1895, an amateur astronomer in Beirut, using a 12-inch refracting telescope, watched fifty to sixty birds in groups of two or three fly south past the disc of the Moon as it hung low on the horizon over a period of a few hours. He was clearly mesmerised by the spectacle. It's likely that these were white storks, flocks of which fly over Lebanon in autumn on their way south. Charles Morton, had he access to a telescope, would probably have interpreted this as further evidence of lunar migration.

The stork is just one migratory bird, but it does seem to have a special connection to the Moon, as well as one to babies. Storks are held to be devoted and attentive parents, and this may be the origin of the widespread story that storks 'deliver' babies (thus erasing the women performing the actual labour), which is told to children to explain the apparently sudden appearance of a newborn. In a 1937 US animation called *The Stork Takes a Holiday*, storks manufacture babies on the far side of the Moon, but apart from that, the source of the connection is very hard to pin down. Nonetheless, images of storks carrying the typical baby bundle across the face of the Moon abound – on baby cards and clothes, pyjamas, T-shirts and graphics. There are echoes of a deep cultural association here.

THE CHILDREN'S MOON

It's not just babies; the Moon plays a special role in the life of small children too. When the Moon is visible in the early evening sky, some say it's the children's Moon, as they're not yet in bed and can see it.

The American psychologist G Stanley Hall (who invited Freud and Jung to the US in 1909) studied children's thoughts extensively. Some time before the turn of the 19th century he conducted a study on what children thought about the Moon. His sample was 423 children, including 321 girls, and the age range was around five to eighteen years old. In 1902, his former student JW Slaughter looked at the data in more detail. I found the children's cosmologies in Slaughter's study, untrammelled by post-war science, a revealing window into a world view where the Moon was a friend, confidante or sometimes judge. There were many accounts of an edible Moon, made of cheese, honey or cream; and many stories of the old man or woman in the Moon. These beliefs are a constant theme of European nursery rhymes and fairy stories. What fascinated me most, though, was how intimately the Moon was woven into the children's lives, not by science but by emotions.

From the ages of three to eight, the sight of the Moon in the night sky energised many children. 'They jump, shout, run, laugh aloud, lose their usual sleepiness, are usually good tempered and often excited to the point of abandon,' said Slaughter. 'The sight of the light may almost intoxicate.' It's not stated in the study but I imagine this as a full Moon.

When low on the horizon, the Moon did not seem far away. Small children thought they could walk there, it seemed so close. And if they got close enough, they could climb into it with a ladder. They had not yet learnt that the Moon was impossibly far away.

Many little kids told the Moon their secrets and troubles, and felt the Moon was kindly and looking after them. Some, though, were afraid the Moon would not be kind if they were naughty. One girl said that she 'talked and shouted a good deal by spells to the man in the moon, and thought he could hear although he did not answer'. How I wish I knew what her spells were!

More recent studies of children's attitudes to the Moon are very different. They test astronomical knowledge — whether the children know what the Moon is made of, how it moves around Earth, and why the phases occur. All very important, I don't doubt, but I prefer to know about the little girl who sang to the Moon and asked it to give her cake and ice cream. That's my kind of Moon.

My first memories of the Moon are the Apollo 11 Moon landing. For almost all of my life, the Moon has been a place where humans have already been for real. They used rockets instead of swans or dew, and it took three days to get there. There was no old Spaniard to greet the astronauts when they arrived.

THE MOON OF SCIENCE OR THE MOON OF LOVERS?

These days, it's easy to forget that the Apollo 11 mission wasn't just about science and politics, but about metaphysics too. People had to come to terms with a new kind of Moon. Many wondered if the eternal mystique of the Moon could survive the onslaught of cold, hard science. What would lovers do if gazing at the night sky made them think of rockets instead of romance? Would humanity lose something precious by taking technology into the realm of myths and legends?

Changing perceptions of the Moon can be mapped through the pages of popular magazines. The *Australian Women's Weekly*, for example, published numerous stories, columns and poetry about the Moon. In the 1930s, poems celebrating the Moon as mysterious and feminine were common. By 1946, after the Second World War, science (and satire) started to creep into the poems. In this excerpt from Australian poet and war correspondent Dorothy Drain, the Man in the Moon laments that he's had to neglect the lovers, poets and songwriters because of increased scientific interest:

> Kindly tell the scientists
> I am overworked
> And wish
> They would leave me alone with my craters.

At this time, the US, the USSR, France and the UK were developing rockets for space launch. The Cold War was heating up and after Earth orbit, the Moon was the next target. The prospect brought science and culture face to face: 'Goodbye, romantic moon,' lamented an unknown writer in the pages of the *Weekly* in 1957. 'Poor lovers: it's black, hot and full of dust.' Lunar science was tarnishing the perfect pearly light, which bestowed ethereal beauty and inspired contemplations of the ineffable.

The kindly Moon beloved of little children was also the confidante of lovers, but this too was threatened by the spectre of Soviet space surveillance. 'For how will lovers be able to gaze uninhibited at the moon, without shrinking into the shrubbery and whispering "Big Brother will be up there any moment"?' asked the 1957 writer. This image recast the Moon as the betrayer of lovers' secrets – if the Russians got there first. And the Russians did get there first. In 1959 the robotic probe Luna 2 became the first human object to land on the Moon, scattering pentagonal medallions stamped with 'USSR 1959' as it crashed.

After the Apollo 8 lunar orbiting mission returned close-up images of the surface in 1968, the *Weekly*'s columnist Robin Adair wrote about how old ideas of the Moon would have to be abandoned:

> As one music magnate remarked, 'What the heck do I do with records and sheet music that go "Blue Moon, I saw you standing alone …"' when Astronaut Lovell said, in effect, 'Moon, I saw you standing below, a whitish grey, like dirty beach sand …?'

With this comparison, 'a thousand poets turned in their graves' according to journalist Kay Keavney.

Adair was scathing of a US protest group called Hands off the Moon, who wished to keep the Moon pristine as a romantic symbol. He admitted, though, 'that it will rather take the mystery and glamour out of the old moon when blokes have roamed around it'.

All of this tongue-in-cheek banter concealed a serious concern. The tension between the Moon of science and the Moon of romance was captured by chemist and writer CP Snow in his influential 1959 essay 'The Two Cultures'. Snow argued that science was in the ascendancy, but that neither the sciences nor the traditional culture of literature and art understood each other, to the detriment of both. There were, he said:

> Literary intellectuals at one pole – at the other
> scientists, and as the most representative, the
> physical scientists. Between the two a gulf of mutual
> incomprehension – sometimes (particularly among
> the young) hostility and dislike, but most of all lack of
> understanding.

In many ways, the Moon was the battleground not only of communist and capitalist Cold War ideologies, but for which of the two cultures would write the script for the universe. By May 1969, the confrontation was imminent. 'In a few weeks', proclaimed a headline in the *Weekly*, 'purple [science] fiction will be changed to prosaic fact when men land on the moon'.

And with one small step, neither the Moon nor Earth was ever the same again.

Was Apollo 11 the end of the lovers' Moon? It seems not. People found a way to reconcile the dissonance between the two cultures in their everyday lives. No doubt many looked up at the Moon in the weeks that followed, wonderingly, and came to the same conclusion as Australian writer Nan Musgrove, who wrote on 6 August 1969:

> It looked so ugly in those pockmarked cratered pictures taken by Apollo 11. But on Sunday night in the sky it was as lovely as ever, caught up in a halo showing through the overcast.

One did not have to choose which Moon to keep; science and the arts had not been fatally riven apart by the encounter with reality. Perhaps, on the contrary, it had brought them closer together.

THE FUTURE OF THE LUNAR PAST

The six Apollo landing sites, however, aren't just about technology or poetry. Along with the Lunas, Rangers, Surveyors, and numerous other missions from China, India, Japan and the European Space Agency – over thirty missions since Luna 2 landed in 1959 – the Moon has become a cultural landscape strewn with archaeological sites. These are unique places: not only are they located on another world, but they demonstrate the changing technology and aspirations of sixty years of lunar exploration, from the Cold War decades when national prestige was the driving force of space, through scientific exploration, to the present day – when commercial exploitation of lunar resources is about to begin.

Naturally, those planning new missions are keen to visit places like Tranquility Base, not just to see the historic traces of those early missions, but to gather evidence of how the lunar environment affects human-manufactured materials that can be used for future planning. The only problem is that the sites are fragile. What we might learn from an archaeological analysis of the astronauts' footprints across the landscape could be erased in a moment by a nearby landing mission stirring up dust, or a rover driving across the site and overlaying its own tracks on the original ones.

In 2011, NASA put together a set of guidelines, with the advice of space archaeologist Beth Laura O'Leary, for protecting the Apollo, Ranger and Surveyor lunar sites from future surface activities that might harm them. These have no legal force, but they set a good precedent for future missions. One of the major concerns was the impact of the abrasive, sticky lunar dust. A key piece of evidence was an extraordinary event that space archaeologist PJ Capelotti calls the first archaeology on the Moon.

Surveyor 3 was part of a series of US robotic landing missions that left seven craft on the surface of the Moon. In November 1969, Apollo 12 landed on the edge of a crater, just 180 metres from Surveyor 3, launched two years earlier. The two astronauts, Pete Conrad and Alan Bean, walked over to the Surveyor 3 and removed a camera and a couple of other pieces to take back to Earth.

Their visit created a completely new type of space site combining human and robotic traces. You could argue that Surveyor 3 and Apollo 12 are now part of the same site as they preserve the evidence of interaction between the two locations. So, unlike all other lunar landing sites, the Surveyor 3 site is the only one which is the result of multiple visits. This rare human-robot encounter on a far-away world is what archaeologists call a 'multiphase occupation'. Capelotti argues that this interaction makes the two sites into an archaeological precinct with very high scientific significance.

The pieces of Surveyor 3 which were returned to Earth were even more interesting. When they were analysed, they were found to have been effectively sandblasted – once from Surveyor's own landing, and again when Apollo 12 landed on the edge of the crater. A number of microscopic pits were observed. Some were probably the result of the micrometeorite bombardment the Moon is constantly subject to with no atmosphere to protect it. Many of these pits were on the side of the Surveyor facing the Lunar Module. It is likely that these were sand-blasted from dust that was blown away from the Apollo landing site by rocket exhaust. Some darkening of painted surfaces due to the effects of solar radiation was also observed. This was very valuable information for working out how more constant traffic on the Moon might affect human habitats or industries, but it also told us what might happen to the old sites if vehicles landed near them.

The Surveyor 3 sample makes this site complex high on the target list for revisiting in a future crewed mission. The analysis provided baseline data about lunar effects on spacecraft materials; and a sample for comparison over fifty years later would be invaluable for science. But how much damage would such a visit cause?

For a long time, as lunar archaeologist Beth Laura O'Leary notes, the lunar sites were protected by their remoteness. This would change when the number of future missions escalated. Surveyor 3 demonstrated what the impacts of stirring up dust would be. Hence one of NASA's recommendations was to create buffer zones around all of the former landing sites so that people or machines on the surface, or rockets and other vehicles flying overhead, minimised the amount of dust movement. It's a reasonable strategy, and one that I've often used myself to protect heritage sites on Earth. When undertaking construction activities, an exclusion zone is created that machinery cannot go inside, anywhere from 10 metres to 1 kilometre. As long as people respect the boundary, the site can be protected from harm. Unfortunately, sometimes the information isn't passed down the line to machinery operators on the ground, or they assume it's not important, so the strategy isn't failsafe.

Its effectiveness also depends on how you define what a 'site' is, a contentious proposition even on Earth. In space, it's even more complex as the Outer Space Treaty and the treaty covering the Moon, the Moon Agreement, forbid making territorial claims in space, and also assign ownership and liability to the state that launched the space object. But for archaeology, this means that the two things we consider to make up a site, the material objects or traces of human activity, and the environment or place they're located in, are legally separated. For example, the US owns all the artefacts from the Apollo missions, but has no legal jurisdiction over the blast zones, footprints, rover tracks and sample collection trenches, which archaeologists would consider essential parts of the site. The 106 objects left at Tranquility Base have been registered with the states of California and New Mexico but the site itself cannot be, because that could be interpreted as making a territorial claim in contravention of the Outer Space Treaty, and that door needs to stay firmly shut. There is no legal protection for the site and as yet, there is no international agreement on how to manage the heritage values of sites which are not on Earth.

This got me thinking about where lunar sites began and ended in relation to the buffer zones. It occurred to me that the buffer zones treated the sites as if they were flat planes on a map, but the places where the Apollo astronauts dug cores and samples out of the lunar regolith had depth, and the landing modules, flags, rovers and experiment packages had height. The sites had a vertical as well as a horizontal dimension. Because of this, they cast shadows. These shadows were of two types: those embedded in the surface within the footprints, furrows and trenches, and those falling on the surface from an object with height. The shadows were not static, following the sun during its two-week cycles. As well as the archaeology of the material culture, each of the sites had a shadow archaeology. This intrigued me as much as the artefacts themselves. Could the shadows be considered part of the fabric of the site?

Shadows on the Moon are, however, of great interest from a completely different perspective. Years ago, I was doing some research about Woomera in the library at the Defence Science and Technology Group headquarters in Adelaide, and found a book that has stuck in my memory. It was a study written for US Congress, in the 1980s, on warfare in space. The chapter on the Moon was all about how lunar conditions would affect hand-to-hand combat. The author, a military expert, noted that many of the properties of lunar light would make such a situation particularly challenging. For example, if you hid from someone in a shadow on Earth, you would still be visible as light scatters and diffuses through our thick atmosphere. This effect is called Rayleigh scattering, and it's what makes the sky appear blue.

On the airless Moon, sunlit areas are very bright and shadows are as black as Hades and as heavy. Although scattered reflection from the dust on the lunar surface will still provide some illumination of objects in shadow, if you go deep enough into one, for example, under a boulder overhang or in a crater, you can be invisible. You're also cold. The temperature in shadow drops markedly.

It was horrifying to me to think that people – presuming the volume I found in the library was just the tip of the iceberg – had been studying planetary environments to assess them for military purposes. But it made me more curious about how shadows differed from those on Earth.

There are places on the Moon that have always been in shadow, perpetually in the darkness. Because of the tilt of the Moon's axis, the rays of the Sun have never reached the bottom of some craters at the poles. The shadows in these locations are over two billion years old and colder than almost anything else in the solar system at minus 249° Celsius (the coldest it's possible to be is −273.15° Celsius, corresponding to zero on the Kelvin scale). Analysis of data from lunar orbiting missions suggests that water ice could be hiding in these deep shadows – a valuable resource for future lunar industry.

AN EPHEMERAL ARCHAEOLOGY

One thing the astronauts from the Apollo missions brought to the Moon was new kinds of shadows, cast by machines and bodies and flags and rovers, in an interplay of movement and stillness. On Earth, the movement of living things; the changing of the environment, both natural and cultural; and the weather, which occludes sunlight to different degrees, makes shadows very dynamic. Lunar shadows, however, are more passive at human time scales, their movement identical with the fortnightly passage of the Sun over the surface.

The Apollo missions brought shadows that were not so passive. The speed of the shadows differed, depending on the activity being carried out, and was much faster than the slow passing of the day. Some shadows were solid black and some were lacy and textured, reflecting the mesh on the umbrella-shaped antennas which adorned the rovers.

They crossed and uncrossed with the angle of the Sun and the movement of the astronauts around the tiny landscapes that constituted their lunar experience. The shadows were captured and frozen in many, many photographs of all the Apollo missions; in these photographs, they became another type of artefact.

And then some shadows left, never to return, and other shadows stayed to be swallowed by the lunar night and to emerge into day again. The shadows of the flags, descent modules, rovers, cameras and other equipment will continue to be cast over the lunar regolith until the objects decay in tens, hundreds or thousands of years. The objects left behind don't move, but their shadows circle them in diurnal devotion, sundials without a mission.

The Lunar Reconnaissance Orbiter (LRO), launched in 2009, used the shadows to detect the presence of these orphaned items of material culture. We've been able to get images of Tranquility Base, and the other landing sites, from the satellite flying over. The famous Apollo 11 flag, alas, is no longer standing, probably knocked over when the ascent module lifted off and blew dust all over the DDE experiment.

In my quest for the meaning of lunar shadows, I first had to investigate what kind of thing a shadow actually was. It was not a silhouette, which is the dark outline of an object, generally against a light background. The silhouette is an image of the thing itself. Nor is it a reflection, where an object throws back light to create a counter-image. It's not albedo, which is the amount of light a surface absorbs.

The more light is absorbed rather than reflected, the darker the surface appears. When we look at the Moon from Earth, much of the difference in light and dark areas results from albedo rather than shadows. It was Sir Isaac Newton, in the 17th century, who established that darkness was not a positive force like light, and shadows were caused by the absence of light.

Then there was chiaroscuro, meaning the contrasts between light and shadows – used by artists to create three-dimensional effects on two-dimensional flat surfaces. Chiaroscuro doesn't just occur on the canvas either: we use shadows in depth perception in everyday life. Shadows themselves are two-dimensional representations of three-dimensional objects.

As I was scrolling through Apollo images in NASA's online archives, I noted the many shadow astronauts with elongated legs, present yet absent in the photographs they were taking for the audience back on Earth. They seemed so lonely and silent, these shadow astronauts; and I was reminded of a well-known series of paintings by the artist Giorgio de Chirico. Around 1910, he started working on a theme in which shadows figured prominently. The paintings were often of empty town squares with statues, towers and arched building façades. The shadows were sharp and elongated and seemed at odds with the strength of the light. He wasn't using the shadows to create depth, but to subvert the conventions of 'perspective illusionism'. I read what I could find about the paintings: all the art historians and critics used words like sinister, foreboding, haunted, desolate, dreamlike, eerie, lonely and sad. These were powerful emotions for such passive shadows to invoke. More than the physical perception, shadows are highly symbolic. They represent melancholy, concealment and secrecy. They symbolise death and the soul, the supernatural, dreams and ghosts, the underworld, coolness and rest on a hot day.

Shadows are visible but not tangible. Their disposition varies with the position of the light sources. They're natural sundials that indicate the passing of time. I wondered if shadows had been a factor in considering the aesthetic values of heritage sites on Earth. My searches in the heritage registers of various countries did not produce much. Usually shadows were considered as part of the built environment, and their impacts were mostly characterised in negative terms, as shadows cast by more recent buildings could affect the perception of a heritage feature on an older building. There were two exceptions I found in the World Heritage List. Both referred to shadows cast by fortified city walls. These shadows created a sense of belonging and protection for the communities living within them. So there was some precedent for considering shadows as part of the fabric and cultural significance of a heritage site.

A DESCENT INTO DARKNESS

It's 21 July 1969. Neil Armstrong is descending the ladder from the Apollo 11 landing module. As he sets foot on the Moon, he utters words which have since become immortal.

But he also says something else. 'It's quite dark here in the shadow and a little hard for me to see that I have good footing.'

He had descended into something the Moon had never seen before: the shadow cast by a crewed vehicle, the landing module. Shadows were to play a large role in the unfolding lunar mission. Landing the Eagle safely meant learning how to interpret what the shadows said about the unevenness of the terrain. Armstrong and Buzz Aldrin had to conduct their scientific work in these unusual shadows, which often made parts of equipment that fell into their deep black invisible. But the shadows were also the subject of investigation. They assessed temperature changes within them, finding their suits buffered them quite well but it was a little cooler. They observed the effects of shadows on visibility.

One of the first things the astronauts did was make observations about the impact their boots had on the lunar regolith. The ridges in the soles, images of which have been reproduced countless times, were in fact an experiment: the contrast between the light and shade in the ridges was a way to measure the reflectance properties of the dust, and the angles allowed calculation of the depth to which the astronauts sank into it. Armstrong commented:

> It does adhere in fine layers like powdered charcoal
> to the sole and sides of my boots. I only go in a small
> fraction of an inch, maybe an eighth of an inch, but I
> can see the footprints of my boots and the treads in the
> fine, sandy particles.

The grey lunar dust around the Apollo 11 site was soon imprinted with the footfalls of the two astronauts as they went about their work, taking samples of soil and rock, setting up experiments and taking photographs. The site is criss-crossed with layers of footprints that built up over the two and a half hours they spent outside.

I'm not the only person who has noted the similarity with another series of famous footprints – but these are from 3.6 million years in Earth's past. They were found by Paul Abell, a geochemist who was a member of Mary Leakey's archaeological team at Laetoli in Tanzania in 1978, nine years after Apollo 11. At this place, three *Australopithecus afarensis* individuals had walked in their bare feet across soft grey volcanic ash, leaving deep, shadowed imprints. The characteristics of the seventy footprints showed a bipedal gait of people with short stature. It was the earliest direct evidence that these human ancestors walked upright. Perhaps the Apollo footprints signal the start of another trajectory, of spacefaring cyborgs.

Over the course of the surface mission, the astronauts took over 600 images and films. In these images we see the long shadows cast by the light of the Sun and Earth around the astronauts, the landing module, flag and other objects. The disposition of the shadows forms part of the corpus of conspiracy theories claiming the lunar landings were faked. It's argued that the angles of the shadows make no sense and are caused by the lighting in a film studio, where actors lumbered about pretending to be in low gravity. This very much reminds me of the counter-intuitive shadows of de Chirico's paintings. People with far greater knowledge than I have debunked the theories, but I think there's another aspect to the shadow conspiracies. Shadows conceal and obfuscate; they create illusions by distorting height and proportions, but Moon conspiracy theorists look to shadows to reveal the truth. In Plato's famous allegory of the cave, captives only know the world through the shadows cast on the cave wall by people and objects moving behind them. The shadows are an imperfect and distorted reflection of the real form of the objects. In their own minds, the conspiracy theorists are like the escaped captive who emerges from the cave to perceive the true cause of the shadows.

The shadows are signs that can be read, and the LRO used them to locate the six Apollo landing sites. Orbiting as close as 50 kilometres from the surface, the LRO could see the shadows cast by the landing modules, instrument packages and even the flags. The astronaut traverses and rover tracks from Apollo 15–17 were visible as dark wiggles, like the burrowings of an insect in tree bark.

There's no doubt in my mind, despite their intangible, ephemeral nature, that the Apollo shadows are a significant part of the fabric of the sites and of their cultural significance. Although abandoned by humans, the shadows mean the sites are not still. They've altered the temperature and light environment that existed in the landscape prior to the landings. It is their difference from Earthly shadows which makes them significant; they are the shadows of humans and human artefacts in the light of another world, and they bring novel geometries and textures to lunar shadow topography.

Beth Laura O'Leary and her team from the Lunar Legacy Project have already catalogued the artefacts at Tranquility Base; I would like to make a catalogue of the shadows. It's not only the hardware and the relationships between objects at the Apollo sites which could be damaged by careless visitation: the chiaroscuro created by the actions of these first humans could also be destroyed.

The shadows I find most compelling, though, are the shadow selfies of absent astronauts stalking on their long legs over the regolith, camera raised to their visor. They feel like the uncanny double or doppelganger, human but not quite. They're silent, lonely and melancholy, as if Tranquility Base were a town square in a de Chirico painting. But there is also a kind of peace. Japanese novelist Junichiro Tanizaki, in his work *In Praise of Shadows*, expresses this perfectly:

> And yet, when we gaze into the darkness that gathers behind the crossbeam, around the flower vase, beneath the shelves, though we know perfectly well it is mere shadow, we are overcome with the feeling that in this small corner of the atmosphere there reigns complete and utter silence; that here in the darkness immutable tranquility holds sway.

SHADOWS AND DUST

The shadows may, however, be of more than passing interest in considering the taphonomy of lunar sites. Brian O'Brien's DDE experiment suggested that dust levitates as the terminator passes over, due to changing electrical potentials. There may be small-scale surface dust movement as shadow boundaries change, creating sub-millimetre electrical fields that cause dust particles to accelerate and collide with each other inside the nooks and crannies of the lunar surface equipment. So the dust environment of the objects left on the Moon may not be static, and we don't know what the long-term impacts on material degradation are.

For a few decades, the lunar sites were safe from disturbance. But now everyone wants to return to the Moon. Distance will no longer protect lunar heritage, if we think it is worth protecting. It's not all about science or prestige any more, either, in the era when wealth is the driving force behind space exploration. The Moon has resources that entrepreneurs on Earth would like to access. These include rare earth elements, like yttrium and ytterbium, which are used in lasers, computers, mobile phones and car batteries; helium-3, which might be used as a clean nuclear fuel, and many others, such as water, which could be used to sustain a colony on the Moon in what's called In Situ Resource Utilisation. The water ice hidden in the permanently shadowed polar craters is a resource that could be used to make fuel, as well as for habitation. People are going to be analysing the shadow landscape to locate resources for future industry.

Private companies have been established to pursue lunar and asteroid mining. Countries like the US and Luxembourg have put in place legislation to support commercial ventures on the Moon and in the asteroid belt. People seem to accept that industry on the Moon is not a matter of *if* any more, but *when*. And when it happens, there is going to be an almighty lot of abrasive, adhesive and corrosive lunar dust stirred up as rockets and rovers come and go. The future of lunar heritage is at risk.

My years of working in the terrestrial mining industry have suddenly become relevant to the future of space exploration. I think off-Earth mining companies are overlooking some critical processes. Disturbing the surface of the Moon at an industrial level could have a negative impact on its cultural significance for the entire population of Earth. This is even before you consider possible damage to the lunar landing sites of many nations. There is an urgent need to develop an environmental management framework for space, and cultural heritage must be part of this. Space archaeologists have an important role to play in the next phase of human engagement with the solar system.

There are innumerable technical problems that need to be solved to have a viable lunar industry, but there's one big one that affects everything we might do on the Moon. The Apollo astronauts found that lunar dust stuck to them and wouldn't come off. It clogged up their equipment seals and caused mechanical equipment to stop working properly. It coated instrument faces so that they couldn't be read – and this was after just a few days. We think of the astronauts in blindingly white spacesuits, but they ended up covered in dirt, much like an archaeologist in the trenches.

The dust contains tiny, sharp spicules of obsidian, a natural glass, that are very abrasive. It's also electrostatically charged from constant bombardment with solar particles and cosmic rays – there's little atmosphere to protect the surface, as we have on Earth. This makes it highly adhesive.

With space vehicles ferrying equipment, personnel and commercial products between the Moon and Earth, there's going to be a lot of dust blown around. If solutions aren't found to control it, it's even possible that it will be blown up into lunar orbit and create a dust cloud around the Moon. This is a critical problem to solve before any industrial activity takes place. John Young, the Apollo 16 commander in 1972, said, 'Dust is the number one concern in returning to the Moon.'

Naturally the dust problem has occupied the minds of scientists working on lunar mining systems a great deal. Proposals to mitigate dust damage include building blast walls to contain the dust, fusing the dust into landing pads so that rocket take-offs and landings don't blow it around, and creating materials that repel the dust to stop it clogging up things. Some of these proposals will also minimise dust abrasion damage on historic lunar spacecraft. This is a rare occasion where the research needed to develop lunar resources also helps us protect some of the most significant sites of the 20th century. So there is some hope that we can ensure that the early history of human adaptation to space environments is not erased.

This still leaves the bigger question. What about the Moon itself? How will people feel if they look at the Moon in the night sky, and know that it is being mined before their eyes? The Moon is a universal cultural symbol that unites us from the earliest human ancestors millions of years ago into the deep future of humanity. So far, all lunar missions have been small-scale and scientific. The presence of human sites, which we can't see from Earth even with the most powerful telescopes, has not diminished the intangible heritage of people's beliefs and dreams. But it might be different when we know that private corporations are making a profit from digging up the Moon.

The reaction of, say, an Australian to a US-based profit-making mine in which they have no say or share could easily be negative. A First Nations Australian may have another layer of reaction, based on their experience of alienation from country and destruction of cultural heritage arising from terrestrial mining. Moreover, an assault on the integrity of a celestial body which belongs to a suite of cultural knowledge in which the past is intimately entwined with the creation of law, identity and land in the present – may be a matter of some concern. Aboriginal people are by no means the only First Nations group to have such a relationship with the Moon.

The Bingham Canyon copper mine, owned by Rio Tinto and located in Utah, is the largest open-cut mine pit in the world. It's been in production since 1906. What started as a tiny pit is now 4 kilometres across, 1900 hectares in area, and 1.2 kilometres deep. Imagine something like this – but on the Moon. How would we feel if we could see it from Earth? We wouldn't be able to see it with the naked eye – as a comparison, the Moon's Tycho crater, which you can see if you choose the right conditions is over 100 kilometres across. It may not be an open pit; strip mining may turn formerly bright regions as dark as the *maria* (the dark areas on the Moon's surface, which early astronomers believed to be seas), extending the waistline of the Man in the Moon. Even if the evidence of mining is not visible to the naked eye, we will see it through satellite imagery and telescopes on the surface of the Earth. There's a very accomplished amateur astronomy community who photograph and post images of the lunar surface all the time. But maybe this visibility will be an advantage. It might be the best way that the people of Earth can monitor and regulate lunar industry.

We shouldn't allow ourselves to be taken by surprise. We should be prepared for changes to the Moon we think we know – to have to make new meanings for it, as we have had to for Earth and other places in the solar system, like Pluto. The Anthropocene era involves redistributing minerals and elements in a way that's geologically visible. If lunar mining goes ahead, there is going to be even more redistribution and exchange of terrestrial and lunar materials. It's a new geology, but it's also an archaeological signature that is going to be distinct on both worlds.

But it won't be the incredible expense, untested technology, undeveloped markets or international treaties that will hinder the exploitation of resources on the Moon. It's more likely to be the microscopic grains of sticky grey dust, a lunar adversary worthy of all our terrestrial science.

THE MANY-COLOURED MOON

The grey dust and stark shadows of the Moon when you're on the surface are very different from the pearl and honey colours, sometimes even shading into the red of a 'blood Moon' when there's a lunar eclipse, that we see from Earth. But Alan Bean, the fourth astronaut to walk on the Moon on the Apollo 12 mission, did not see it as a colourless greyscale landscape. He was a painter; and conscious that he was the only one with this kind of vision to see the Moon first-hand. He went back into space on a Skylab mission in 1973, and retired in 1981 to focus on his painting.

In 2010, he told Dr Phil Metzger, one of the scientists researching how to control lunar dust, that he didn't see the Moon as grey. For him, it was full of colour. In the textures created by the movement of light over the lunar rocks and sediments he saw and painted shades of blue, turquoise, mauve, yellow and brown. This vision was not the product of Newtonian colour physics: the colours evident in the subtle interplay of shadows and dust was more akin to the colour theory of the poet, writer and phenomenologist Johann Wolfgang von Goethe.

Newton's study of light showed that shadows were constituted by its absence. He also pioneered an approach to colour that is the dominant one today, when he split light into the wavelengths of the rainbow by beaming it through two prisms. This wasn't the only way to look at colour, though. Goethe, the celebrated author of *Faust*, brought a poet's perception to the science of light and dark. Goethe's theory regarded shadows as an integral part of how we perceive colour. One of his most radical arguments was a refutation of Newton's ideas about the colour spectrum, suggesting that shadows were an active ingredient in the creation of colour rather than the mere passive absence of light. 'Colour itself,' he wrote in 1810, 'is a degree of darkness ... allied to shadow.' He experimented with making shadows of many colours, noting that this 'may render the most pleasing service to the painter who knows how to make use of it'.

His anti-Newtonian approach earned him a pummelling from the scientists of the day and his colour theory was roundly rejected in favour of Newton's, who was, of course, right. This doesn't mean, however, that there is no value in considering Goethe's phenomenological approach, based more on perception and experience than physics. It seems to me that this is precisely what we need to understand the modern Moon, where perception and meaning may be as illuminating as scientific measurement.

Only twelve people have ever seen the surface of the Moon from inside their own bodies. One day, perhaps in the not too distant future, that number will increase, and perhaps there may be poets and artists among them who will see new colours and textures in the shadows and dust, to refigure the Moon for new generations.

CHAPTER 6

THE EDGE OF
KNOWN SPACE

Black upon black, the fissure in the ice,
The outer rim where you passed
Once, but not twice.

AC Gorman, 'Eurydyssey'

It's extraordinary when you think about it. In 1972, a mere fifteen years after the first human object left the atmosphere, the first deep space probe was sent to explore the far reaches of the solar system and to continue beyond into interstellar space. Pioneer 10 is still out there, with its twin Pioneer 11. Their present locations are just predicted trajectories; we have no way of knowing their exact position, and we can't communicate with them.

Both Pioneers carry a gold-anodised aluminium plaque designed by astronomers Frank Drake and Carl Sagan, with artist and writer Linda Sagan, intended to carry a message to whoever might find the spacecraft. There's a bunch of scientific symbols, including the location of Earth, but the image people talk about is the line drawing of a naked man and woman. The man stands with his hand raised as if in greeting (it was also to show the opposable thumb) and the long-haired woman stands nonchalantly with her weight on one leg. The man looks straight out at you, but the woman looks demurely to one side in the man's direction. Their general appearance is based on the drawings of Leonardo da Vinci and classical Greek sculptures.

Hypothetically, will anyone be able to interpret them 30 000 years into the future? The human figures may be as opaque as the painted animal plaques from the Apollo 11 Cave, 30 000 years in the past. Archaeologists puzzle over whether the funny-looking feline-antelope thing is a therianthrope, combining human and animal characteristics. This is because the animal's hind legs, which have been painted over an original pair, appear to be human. There are no other examples to compare with from this period, so it's hard to know if this was typical or an anomaly. The hypothetical alien interceptors of Pioneer 10 and 11 will have just this one example of engraved human art; they won't have anything to compare it with either.

One of the common interpretations of archaeological rock art animals is 'increase' magic, in relation to an animal that is hunted as a food source. The purpose of painting the animal is to ensure its numbers continue to increase, or perhaps to bring good luck in hunting. If we applied that logic to the Pioneer plaques, the human figures might be assumed to be prey – and an invitation to come to dinner.

There are all kinds of things that rock art might mean, but without being part of the culture that made the art, we have no way of knowing. It's pretty much the same for all our attempts to send messages out into the void for others to read. The rock surface has simply been replaced by a metal plaque and left in a different location. Future aliens, or past humans: the problems of trying to communicate through symbols are equally challenging.

Let's try another comparison out for size. One of the most famous ancient art works is the Venus of Willendorf. The 11-centimetre-tall limestone figurine was excavated from the Austrian site of Willendorf in 1908 and she's also about 30 000 years old. Without facial features, or even feet, she's all sexuality, with great breasts and belly and a well-defined vulva. A multitude of theories have swirled around her: mother goddess religions, matriarchal cultures, Palaeolithic pornography, prehistoric selfie, the power of post-menopausal women, cross-cultural communication. She confounds interpretation and this is one of the many reasons I love her.

Because of the emphasis on her secondary sexual characteristics, some have proposed that she was made for Palaeolithic male viewing pleasure. (This is not the consensus interpretation of the Venus. Personally, I prefer the 'grandmother hypothesis', which acknowledges the important contributions of older women to society.) By analogy, if our future aliens have any conception that the Pioneer drawings are bodies, they could equally interpret them as porn – which was, it seems, an actual consideration when the plaques were being designed. The big controversy is about Pioneer woman's missing 'pudendal cleft'. The man has a nice discreet package, but she's featureless – a kind of reverse Venus of Willendorf, who gets to have a face and feet instead. (Neither Pioneer figure has pubic or body hair.) The line was there originally, but had to be erased before NASA would approve the image. US standards of 'obscenity' were being imposed on aliens we didn't even know would read the message! There's a rock art precedent for this too: any shape with a cleft in it – a line that starts on one side of the shape but doesn't continue to meet the other side – is often presumed to be a vulva. Nudity is in the eye of the beholder, as art critic John Berger argued in *Ways of Seeing*.

The naked couple accompanied Pioneer 10 on its flyby of Jupiter, the first spacecraft to draw near the gas giant with its red eye. Pioneer 11, launched in 1973, went to Jupiter and on to Saturn to see the rings in all their marvellous symmetry. After their planetary missions, both spacecraft sailed onwards, gathering speed, and finally were set on a trajectory to escape the solar system, carrying the new Adam and Eve with them.

THE NEW WORLDS

I don't remember any longer what it was like in those days, when planets beyond Mars were all mysterious bodies only ever seen through telescopes. Over my lifetime, we've gone from relying on optical and radio telescopes located on Earth to tell us about the worlds outside our world, to becoming intimate with planets and moons from our own backyard to Pluto. We've travelled similar distances into the human past over these decades too. The Time-Life book on human evolution I used to pore over at my uncle John's house, with its 'March of Progress' illustration showing different hominin species gradually unbending to walk upright, is ridiculously outdated now. Carbon dating, which revolutionised archaeology in the 1950s, has been joined by methods such as thermoluminescence and uranium series to provide older and more accurate age determinations for human remains and sites. In the 1960s, Mary Leakey showed that the human ancestor *Homo habilis* had been making stone tools at 1.75 million years ago. Now we have the 3.3 million-year-old stone tools from the site of Lomekwi in Kenya – and we don't know who made them. Previously unknown human species, like *Homo floresiensis* (known as the Hobbit) have been identified, and thanks to DNA analysis, the much-maligned Neanderthals now turn out to be our ancestors rather than an evolutionary failure. From the 1960s until the present, the scientific evidence for the Aboriginal occupation of Australia has gone from 19 000 years to 65 000 years, and likely older. Looking down has been as much of a journey in understanding our place in the cosmos as looking up.

There have been flybys, probes and orbiters to most planets in the solar system, as well as a few asteroids and comets. We're not doing too badly in the inner and middle solar system. Throughout the 1960s the Moon became as familiar as the neighbour's holiday snaps, as successive missions showed us more and more of its surface. In 1965, it was the turn of Mars as the US Mariner 4 'flew by'.

Mariner 4 reminds me of an inquisitive fly. Its chunky octagonal body, 127 centimetres broad, with various antennas sticking up from it, had a single camera eye, and four 'wings' – solar panels for power, with jointed solar vanes hinged at the ends. I can almost hear it buzzing as it darts hither and thither. Its journey, of course, was more of a slow sail, with the vanes at the end of the wings twitching every now and then to stabilise the craft, as it focused its star tracker on Canopus, the second brightest star in the night sky. Mariner 4 was the first spacecraft to navigate by the stars; all previous missions to the Moon and Venus used the solar system planets for navigation.

Mariner 4 took pictures of Mars when it was a mere 9846 kilometres away. The images were transmitted bit by bit, line by patient line. Overnight on 16 July in 1965, US time, the first forty lines came through, received at the Deep Space Network antennas at Johannesburg, Woomera and Tidbinbilla. The big news was that the images had arrived successfully and appeared to be of Mars, rather than the ceiling, or the inside of the camera housing. The news was conveyed from Woomera to Tidbinbilla in code, accompanied by 'one word spoken laconically and in a broad Australian accent: Hooray!' The Jet Propulsion Laboratory team waited for hours for each line to arrive, slowly building up the closest picture of the planet ever seen. The twenty-two images would take four days to arrive. It's a trite comparison to make, but just imagine if we had to wait that long for data now!

The images were startling. There were no canals showing the water management practices of a dying civilisation; no cities or pyramids or Oyarsas, the benevolent celestial rulers imagined by CS Lewis. The elegant Queen Aelita, subject of a 1924 Soviet silent film, was not stalking the staircases of Mars in geometric Art Deco costumes. What they saw was an old cratered surface that resembled the Moon more than anything else. It was exciting – but ultimately disappointing. If life on Mars existed, it wasn't running around in the open waiting to be discovered. It was likely to be 'simple', or even extinct.

Mariners 6 and 7 followed in 1969; they sent back more images of the cratered regions, still in black and white like the Mariner 4 images. In November 1971, Mariner 9, which looked more like a ceiling fan with its sub-conical body and four solar panels, entered Mars orbit, just two weeks before the USSR's Mars 2 and 3. (There is a lot of Soviet material culture on the surface of the Moon, Mars and Venus, a fact often forgotten.) There was only one hitch: a planet-wide dust storm that obscured the surface. When the storm subsided, a different Mars was revealed: ancient river beds, mountains, deep canyons, and the largest volcano in the solar system, Olympus Mons.

The real showstopper was Viking 1 in 1975. The mission successfully deployed an orbiter and a lander which transmitted significant amounts of data for 2245 Martian sols (a sol is slightly longer than an Earth day). The hope was that being on the Martian surface might reveal signs of life that the earlier Mariners had missed. At last we saw the Martian surface in its true glory, the red planet we always knew it would be. Viking 1's first colour transmission showed a blush-pink sky over a deep red rock-strewn field, taken at Martian high noon. It looks like a warm desert; a tumbleweed or bleached cow's skull would not be out of place. It feels like water or vegetation are just over the horizon, out of view. It's a struggle to comprehend that no matter how far you travel, the gullies are all dry. There will never be a trickle of water or a hardy saltbush. The land is empty and silent but for the wind. Perhaps in this silence, we could hear things no longer audible on Earth: the deep creaking of the bedrock and tremors from the molten iron core.

Over time, our sensory impressions of space have expanded. The palette of the solar system outside the blue Earth has grown to include suites of blacks, greys and reds which require new pages in the archaeologist's colour bible, the *Munsell Book of Color*. Martian reds are familiar now as we've followed the journeys of rovers like Sojourner (over three months in 1997) and Curiosity (which landed in 2012 and is still going strong). The corollary of all this familiarity was, however, that as planets and moons of the solar system revealed themselves, the hopes of finding sentient companions dwindled.

The same process happened for Venus. The enticing vision of a Venus teeming with life was dented by NASA's Mariner 2 flyby mission in 1962, and destroyed by data from the USSR's Venera missions from the 1960s to the 1980s. Sadly, there was not an ocean in sight. There were no swimming V-frogs, as Isaac Asimov imagined in his young adult novel *Lucky Starr and the Oceans of Venus* in 1954. I still love this novel for the sense it gives of just how quickly our knowledge of the solar system was expanding. For later editions, such as the one I read in the late 1970s, Asimov had to write a preface explaining why his Venus was so different from the one people knew about from the Venera missions. In twenty years the memory of oceanic Venus had faded, replaced by a hot, dry, inhospitable world with orange sulphuric acid clouds. It makes me long to have experienced the days before I was born when Earthlings could dream of wonders in the rest of the solar system, before we found out that we were effectively alone. And the more we found out, the more alone we became. There's always hope, though, and in recent years, more and more water has been detected across the solar system. Our neighbours might be microbial in the end, but they will still revolutionise how we think of life on Earth.

Several deep space missions have gone far past their target planets and onwards into even more strange regions: the Kuiper Belt, home of dwarf planets and full of icy rocks; the heliopause where the solar wind stops; and finally, interstellar space. This is where the influence of the Sun fades away, and it's held to be where the solar system ends. Out there, in the cold and dark, so far from home, are the lost Pioneer 10 and 11. They're exactly the sort of artefact that humans might hope to find in the search for extraterrestrial life, but which no-one will ever find, except by the biggest galactic fluke ever. And the Pioneers are silent little shadows with nothing to give their presence away to an observer except an alpha particle popped off every now and then from their plutonium-238 RTGs. Little dead skeletons, like plankton drifting in a deep ocean.

The not-yet-lost Voyagers 1 and 2 might have some hope of being detected even after their power runs out. I've always loved the Voyagers, because they have it all, really: a mixtape for the cosmos, a mission to an unimaginable place, a window into a future that doesn't even involve humans. I'm also fascinated by how humans relate to them. We find ways to make them personal, to draw the far-away and the impossibly scientific closer to us. We give the spacecraft personalities; we make them our avatars. They are proxies for what we long to be, eternal citizens of the vacuum.

Both spacecraft have the next generation of message for aliens in the form of the 'Golden Records', which contain ninety minutes of music, sounds and voices, as well as 116 images (there is a naked couple among them, but they're in silhouette this time, no need for the alien lads to get hot and bothered about pudendal clefts). The Golden Records are coated in uranium-238, which has a half-life of 4.5 billion years, compared to the 87.7 years for plutonium-238 of the RTG power sources. The coating was a very deliberate attempt to give future alien archaeologists – perhaps those that live in the galactic neighbourhood of the star Gliese 445 which Voyager 1 will pass in about 40 000 years from now – a way of dating the spacecraft. The radioactive isotope uranium-238 decays into a series of unstable isotopes, which further decay until they've all turned into the stable lead-206. The proportions of the different elements can be measured to work out how much time has elapsed. What this means is that billions of years from now, the Golden Records will no longer shine as the coating will be the dull grey of lead, but this conversion will be the key to their date and the distance they've travelled.

It's obvious, of course, but I want to say it all the same: the further away you are from Earth, the fewer human arte-facts there are. To the outside observer, it might not be clear that these spacecraft are the forerunners – we hope – of a time when there are many spacecraft exploring the outer solar system, perhaps even visiting human settlements. At the moment, I fear they might look like the end of some-thing instead: time capsules of human culture fleeing the solar system to escape some cataclysmic disaster.

The science of deep space exploration is fascinating but I'm more interested in how we came to know the solar system and how we find it meaningful. What these tiny spacecraft mean is that the entire solar system is a human place. Our senses, through these robotic avatars, have reached into places we can't go ourselves. We have used the physical bodies of the spacecraft to imbue space with human meaning – and human culture. It's all out there, if we allow ourselves to send a little gossamer thread of thought through the solar system and feel the spaces experienced by machine ears and eyes.

THE ARCHAEOLOGY OF NOT-QUITE-THERE

It's interesting to compare the difference between the achingly slow downlink of Mariner 4's pictures of Mars in 1965 with the New Horizons spacecraft's more recent flyby of Pluto, the furthest 'planet' from the sun, in 2015. Pluto's demotion to the status of dwarf planet in 2006 has been debated endlessly. I'm not taking sides, but I love that people care about Pluto. They feel so strongly about it that Michael Brown, the astronomer largely responsible for the reclassification, is known as the 'Pluto Killer'.

Ironically, the same year Pluto the planet 'died' was the year that a spacecraft the size of an industrial fridge was launched on a long journey to rendezvous with it. The only images of the planet and its moon Charon prior to this were some rough pixelated discs which looked like they were designed to hide the face of a witness in a criminal investigation. Like every other space fan, I was enthralled by New Horizons' approach to Pluto. The mission caught the imagination of the world for a couple of epic weeks in July 2015. Every hour the spacecraft drew closer was filled with drama and suspense as the public waited to see what wonders would be revealed. The raw images came in and were processed quickly – it was exciting waking up every morning and logging in to see what new alien landscape had been uploaded overnight.

Something about the flyby put me in mind of an extraordinary poem by Hilda Doolittle, known as HD. In Greek mythology, Orpheus tries to rescue his dead wife Eurydice from the underworld, but makes a fatal error – he looks back at her before she has fully emerged into the light of upper Earth, and she sinks back into the darkness forever. The stark repetitive language of the poem cuts you to the core. It speaks directly to the encounter of a planet and a robot in the outer darkness: 'If you had let me rest with the dead,/ I had forgot you/ and the past'. The robot looks, turns away, and abandons the planet to its solitude.

As New Horizons continued its journey beyond the formerly elusive and controversial planet and into the Kuiper Belt to meet Ultima Thule, I thought about how this encounter could be framed as a kind of archaeological survey.

This mission turned Pluto from a fuzzy disc of reflected light into a place that we can see and read with our own signs and meanings. There was such a wide range of interpretations in the early days of the flyby. Some were geological – craters, valleys, mountains. Others were geometric: rectangles, lines, circles, shapes, smoothness, spikiness. Then there were the zoomorphs: a whale's tail, the head of the Disney cartoon dog Pluto. (Incidentally, the Pluto character was likely named to capitalise on the sensation caused by the discovery of Pluto the planet in 1930.) Perhaps the most charming was that which saw a heart outlined by the large icefields now called Tombaugh Regio, after the astronomer Clyde Tombaugh who first discovered Pluto.

We could also look at Pluto from a heritage/landscape perspective. Using the World Heritage Convention's definitions of cultural landscapes, you could argue that it is an associative landscape: it has no actual physical human traces, but remains a repository of beliefs, dreams and visions. Some of these relate to its location so far from the light of the Sun. It's said that midday on Pluto is as bright as Earth after sunset. It is a cold, dim planet, like a deep sea fish that lives and swims in darkness all its life. Since Pluto was given the name of the ancient Roman god of the underworld, suggested by 11-year-old Venetia Burney in the same year as its discovery, the International Astronomical Union has decreed that place names on it will relate to this theme (which includes exploration). So already we have a set of metaphors and associations that shape the planet in our minds.

Our gaze creates Pluto as a place. Assigning names to features on Pluto is a colonial process of mastery that draws the dwarf planet into our geopolitical web. Despite the efforts of the International Astronomical Union to diversify the cultures represented in named things across the solar system, the old habits die hard. When telemetry showed that New Horizons had successfully completed its flyby, many in the mission control room waved American flags. As I watched the flag-waving, I thought that for some people at least, the passage of a piece of space hardware has been the equivalent of planting a flag – as it was on the Moon – to stamp the outer solar system as part of an American colonial landscape. Indeed, one of the 'cultural artefacts' on board the spacecraft was a US flag, in stark contrast to the universal symbols on the Pioneer and Voyager spacecraft.

The colonial narrative was balanced by the discourse of space exploration as a shared global human endeavour. There was undeniable excitement generated in social and traditional media across nations and age groups, not to mention through NASA's brilliant open access data, with all its materials available to anyone with an internet connection. Possibly more people than ever before waited with bated breath for the experience of gazing on the face of another planet for the first time, like a distant relative they'd longed to meet.

Pluto is a 'natural' object that existed for billions of years before there were sentient beings to observe it. But since its existence was first hypothesised in the 19th century, it has become a cultural artefact too. Like an archaeological excavation, the cameras of New Horizons excavated it from obscurity in the darkness of the night sky and brought it into the light. Space scientists and the public speculated about this strange new artefact: what was it made of? How had its surface markings been created? Its composition, history and appearance were compared to other planets and moons to gauge its character, looking for clues about the genesis of the landscapes captured by the cameras. Deep time was being decoded from a surface of mountains, ice plains and craters: four dimensions being raised up from two.

The images, received as radio signals, converted and processed into pictures that we can view and understand, are a type of heritage or archaeological object in themselves. In one sense they are ephemeral, composed of numbers inside computers. In another sense, their endless duplication and proliferation as they are downloaded and viewed by millions of people gives them a digital resilience. Pluto, in our homes and at work, is a place we visit through a flat screen, almost as if we were inside our own spacecraft.

It strikes me that a lot of planetary exploration is about being not quite, but nearly there. Many planets and moons in the solar system have only ever been 'flown past': a fleeting glance, like when you watch the landscape whizzing past from inside a train carriage. It is very like Plato's allegory of the cave, trying to work out the nature of another planetary reality from shadows and dust. Similarly, there are places on Earth, like the deep sea, that we know only from remote sensing, probes and flyovers. How do you analyse something you can't touch? It doesn't bother planetary scientists who rely on data acquired from remote sensing. This is what characterises space archaeology too, at least at the present time. It's all an archaeology of 'not-quite-there', where we use historical and proxy data in order to make hypotheses about what lies beyond our reach in space.

It does seem counter-intuitive, though, to look at places where humans have never been and where they could not survive, as cultural landscapes. The human dimensions of teacups and tables, the small artefacts of everyday life, don't seem to play any part in this. This is perhaps why the pyramids, and other such wonders of the ancient world, seem so extraordinary and even, to some, alien. Their scale seems to defy human capacity. In the modern industrial world, there are many such structures: bridges, buildings, mines, rockets, giant offshore gas rigs. Of these objects, archaeologist Matt Edgeworth notes that 'The sheer scale ... is such that it is impossible to fully grasp either in a visual or tactile sense – to see all of it at once or to touch more than a tiny part of it.'

John Schofield, an archaeologist who studies the contemporary past, and physicist David Jenkins, have written about the vast differences in scale between sub-atomic particles and the deep galaxy maps that almost show the beginnings of time, all coalesced around the CERN research facility on the border of France and Switzerland. This massive complex of buildings and installations, where the highest of high science takes place (such as the confirmation of the existence of the Higgs boson in 2012) also contains dilapidated chairs and shabby rooms, forgotten corners where Schofield and Jenkins found the empty space resonating with human absence.

I'm left with the feeling that we're becoming very good at bridging these differences of scale. Spacecraft like Voyager and New Horizons force us to get our heads around dimensions of both time and space that are beyond the terrestrial. New Horizons presented the act of viewing, through the feeble jellies of our eyes, as a temporal act. We drew closer and closer to Pluto's surface, like the process of focusing a microscope objective, and when the object precipitated into our field of view, we assessed, named, categorised, analysed – and dreamt. Pluto became part of our personal solar system.

THE GHOST IN THE MACHINE

After its flyby of Ultima Thule, New Horizons is currently heading deeper into the Kuiper Belt to investigate another of the small, icy, rocky bodies that orbit beyond Pluto. Its instruments will function until around 2026. Around 2040, it should enter interstellar space, joining the four previous spacecraft that have also left the Sun behind. Pioneer 10 and 11 might be lost, but we are still in touch with the Voyager 1 and 2 deep space probes.

The Voyagers were designed and made by NASA's Jet Propulsion Laboratory (JPL) in California. Voyager 2 was launched on 20 August 1977. (Confusingly, Voyager 1 was actually launched after Voyager 2, on 5 September 1977.) The 'Golden Records' on them are identical, however, and there's a little bit of Australia on there: a picture of the Sydney Opera House, still under construction with scaffolding around the base, two Aboriginal songs, and an Australian voice speaking a greeting. Only he's speaking in Esperanto, a constructed language invented in the 19th century intended to serve as a global lingua franca. Ralph L Harry, the Australian ambassador to the United Nations at the time, was an Esperanto evangelist. His message was *Ni strebas vivi en paco kun la popoloj de la tuta mondo, el la tuta Kosmo* (We strive to live in peace with the peoples of the whole world, of the whole cosmos).

The primary mission of both spacecraft was to fly by Jupiter and Saturn, following in the footsteps of the Pioneers. After that, Voyager 1 continued towards interstellar space while Voyager 2 went on to Uranus and Neptune: it is the only spacecraft to have taken images of these distant planets. After Voyager 2 passed Neptune in 1989, both Voyagers were officially on the Voyager Interstellar Mission (VIM), to find out what it's like outside the solar system in the stuff between the stars.

In 2007, Voyager 2 passed the termination shock, where the solar wind slows down as it starts to interact with the interstellar wind. The latest evidence from its sensors in late 2018 suggests that it too may have entered interstellar space. Voyager 1 is close to 20 light hours away from Earth and Voyager 2 is over 16 light hours away, or 17 556 000 000 kilometres. The spacecraft travel 1.3 million kilometres each day. It's pretty extraordinary to think that the Voyagers experience the wind from stars other than the Sun. Voyager 2 will be the one to tell us about that, as only it has the instrument to measure the wind particles.

For decades, most of everything we knew about the middle and outer solar system – the gas and ice giants of Jupiter, Saturn, Uranus and Neptune – came from the Voyager flybys. They found previously unknown moons around every planet they visited. Their data allowed the mass of the planets to be calculated more precisely. And there were totally unexpected things like volcanoes on Io, and Saturn's rings being kinked. Always, the planets are full of surprises.

Voyager 2 is the only spacecraft to have visited both Uranus and Neptune. Compared to Jupiter with its spectacular coloured storms, and the stark geometric beauty of Saturn, these planets almost look bland in the optical range of Newton's rainbow spectrum. As it happens most of the action is above or below the wavelengths the human eye can perceive, a good illustration that our visual way of knowing the solar system does not capture everything. So far from the Sun, you'd naturally expect these planets to be on the cool side; but Voyager 2 found that they were warmer than predicted. With much less sunlight, there should be less energy to drive wind and storm systems, but when Voyager 2 measured Neptune's winds, they were over 2000 kilometres per hour, much faster than even those on Jupiter, the stormiest planet in the solar system.

The Voyagers would have 'felt' the influence of the Sun diminishing the further they travelled away from it. In the early years, the warmth of the Sun was beating on their backs with the solar wind, the particles that stream constantly outwards from the Sun, wafting them forward. Every now and then there'd be a coronal mass ejection, when the Sun flings out huge masses of ionised gases called plasma, and the spacecraft would be bombarded with higher energy particles. Gradually, the Sun would have shrunk in size to become just a bright dot; and Earth itself, as we know because Carl Sagan persuaded NASA to turn Voyager 1's camera back towards the Sun one last time, just a pale blue dot. Outside the solar system, the spacecraft no longer feel the coronal mass ejections. The gravitational drag of the Sun is perceptible but too weak to drag the spacecraft back into orbit. And the spacecraft are cold, so cold at 20 Kelvin (corresponding to 250° Celsius). When the Sun sensors get turned off to conserve power, the Sun will cease to exist for them. Eventually, it will be just one of a billion dots of light in the arms of the galaxy. There's no way home from there.

For a long while we got pictures from the Voyagers, but their cameras were turned off in 1990 to save power. As available electrical power decreases, the spacecraft will not have enough to run all their instruments at the same time. Starting in 2020, JPL will have to choose which of the current instruments to keep running. Around 2025 there won't be enough to run even a single instrument. There'll be enough power to send radio signals but that's about it. It will be a slow death, like gradual organ failure.

It's very important to me to know that we are in contact with the Voyagers. Obviously I'm not hanging out at the Canberra Deep Space Communication Complex every day watching the data. But every time a new command is sent to Voyager 2, the amazing Paul Filmer of the US National Science Foundation tweets it from the spacecraft's unofficial account. Periodically @NSFVoyager2 gives the distance from Earth in light years and astronomical units. A few years ago, the account was out of commission for a while and I experienced something quite peculiar. Knowing I couldn't hear Voyager 2's current status made me feel like I had lost a sense, as if the solar system had suddenly contracted inwards. I felt as if my ocean had become a fishbowl. I wasn't aware of Voyager 2 most of the time, but when threatened with its loss, I found myself grieving. Fortunately, @NSFVoyager2 came back online and saved me from existential crisis. But I'd had a taste of how I'll feel when the Voyagers are no longer transmitting and New Horizons becomes the last human outpost in the solar system. For a little while, at least.

If the worst-case Kessler Syndrome happens and no more spacecraft venture out from Earth, imagine how it will be as those that are left slowly fall silent. Bits of the solar system will drop out of our ken like phantom limbs we can no longer feel or flex. It might take decades, maybe a century or more before there's nothing transmitting out there that we can hear. In the meantime, tracking antennas on Earth will be allowed to decay and no new generations of satellite engineers will be trained to maintain them. I can't help but feel this would be a stagnation and a bitter one at that. Failure to develop more sustainable ways of being spacefaring has consequences.

THE PLACE DEFINED BY WIND

Voyager 1 is in a most peculiar place, where the solar wind and the interstellar wind meet and mingle. The edge of the solar system is where the Sun blows a bubble with its winds, called the heliosphere, around all the planets, pushing the slightly denser and colder interstellar medium away. It's not a place you can see using human senses, and what we know about it comes from the observations of Earth-based telescopes and Voyager 1's instruments. There are no pictures because the cameras had been turned off more than a decade before Voyager 1 got there.

It's remarkably hard to describe what this place is like,

but I'm going to try. The average outside temperature is likely to be somewhere between 10 and 20 degrees above absolute zero (−263° to −253° Celsius), a molecule-numbing full 165 degrees colder than the coldest temperature ever recorded on Earth (−98° Celsius in Antarctica, measured by satellite data in 2018). Other parts of the interstellar medium, where new stars are forming, are thousands of degrees hot, but this isn't happening out on the edges of our solar system.

Gases are 99 per cent of the interstellar medium, and there are three types out there. One is gas as we know it, composed of molecules. Oxygen gas on Earth, for example, occurs in pairs of atoms bound together which is written as O_2. There is some oxygen in this region of the interstellar medium, but by far the most common element, as it is throughout the universe, is hydrogen. This crucial fact we owe to the astronomer Cecilia Payne-Gaposchkin. Before she started studying stars in the 1920s, astronomers believed that the elemental composition of the Sun and other stars was pretty similar to that of Earth's crust. Hydrogen and oxygen occur in molecular form, but also atomic form, just one hydrogen or oxygen atom by itself. Finally, there are ionised gases, called plasmas, where the electrons have been stripped away from the atomic nucleus by high thermal kinetic energy and both the charged electrons and nuclei are roiling around together. These come from the Sun. They blow outwards at 400 kilometres per second, but as the wind reaches the interstellar medium, it slows and the plasma heats up as the ionised particles get compressed against each other in an area called the termination shock. Both Voyagers passed the termination shock safely.

Out in this region, called the heliosheath, there are abundant cosmic rays, high-energy particles moving close to the speed of light. They're usually protons, but also include electrons and atomic nuclei composed of both. They come from supernova explosions of stars in our galaxy and others. Only some penetrate the heliosphere, and even fewer reach Earth and make it through the atmosphere – but it's still a fair number, about 10 000 per second. (The Moon is unprotected by atmosphere, however; and when these cosmic rays strike the surface, they cause a tiny nuclear reaction. This probably isn't going to help the preservation of lunar heritage sites.) Voyager 1 isn't protected at all, and is exposed to radiation far beyond that which spacecraft inside the solar system experience.

And there's dust too, about 1 per cent of the particles. Much much finer than terrestrial dust, these particles are mainly made of carbon, silicon and oxygen. We know what they look like because Earth moves through a fine rain of similar dust, left over from the formation of the solar system 4.6 billion years ago. Some makes it down to the surface where scientists look for it to study. Under a microscope the grains betray their origin by their sphericity, a result of melting in their journey through the atmosphere, and feather-like crystal growths. Great numbers of them are preserved in Antarctic ice cores; but if you know what you're looking for, you can find these spacefaring grains in the gutters of your own house.

Voyager 1 has passed through many types of dust. On Earth, it was constructed inside a clean room, designed to minimise the amount of terrestrial dust and other contaminants that might interfere with the satellite's function. On its more-than-forty-year voyage into interstellar space it's been moving through incoming cosmic dust and interplanetary dust from asteroids, comets and the formation of the solar system. The spacecraft's nooks and crannies may have an accumulation of dust in microscopic layers which can be peeled back to reconstruct the journey from Earth, just like an archaeological excavation. Perhaps the aliens who find Voyager 1 will be far more interested in what the dust accumulation can tell them about conditions in our solar system than the technology or culture of the spacecraft itself. At the bottom of the 'trench' will be the most precious dust grains: minute windows into the geology of Earth.

Voyager has made the idea of the edge of the solar system something concrete and tangible, a place we can speculate about and almost feel through the limited senses the spacecraft use to measure it. The Pioneer plaques and Voyager Golden Records imprint this place with human meaning which flows both forwards and backwards between the past on Earth and the impossibly far future. How humans interpret the symbolism changes with every decade that passes. The plaques and records are frozen snapshots of a particular moment in time, a Pompeii in space, while terrestrial cultures move on. But symbols are resilient, like the Venus of Willendorf, speaking to descendent generations across thousands of years.

BEYOND THE MORNING STAR

'Surely the Golden Records are more important than the dust!', I hear you cry. I'm not so sure about that. Records are old technology now even for humans; despite a recent resurgence in the popularity of vinyl, I have my doubts about whether most millennials would be able to correctly interpret the playing instructions engraved on the outside of the records, let alone an alien. But this may not matter – in many ways, the Golden Records were always more about how one particular spacefaring culture saw itself, than a message to unknown species.

The Voyager Golden Records were designed to encapsulate the aural heritage of Earth in 90 minutes. When I realised they included Aboriginal music, I knew I had to find out the story behind it. Recognising the names and identities of Indigenous people when they have been erased in the historical record is an important, though small, act of reversing the effects of colonialism. The trail I followed led to the names of the musicians and an unexpected discovery: that the music in the official Voyager tracklist was incorrect.

As the Voyager missions approached their launch window in 1977, American astronomer Carl Sagan put together a committee to discuss a 'time capsule' for the interstellar component of the mission. Astronomer Frank Drake suggested a record rather than a plaque, as was used on the earlier Pioneers 10 and 11, and suddenly, music was on the table.

The process of selecting this 'world music' is described

in Sagan's book *Murmurs of Earth*. Making the final cut depended on the quality of the recording, cultural diversity, and geographic and chronological range. Sagan was also influenced by an ethnomusicological theory called cantometrics, which used the new computing power of the 1960s to statistically analyse the music of the world, correlating different musical styles with social structures. The hope was that the records could not only represent the diversity of human culture but its evolution as well. The records would be an aural excavation of the human sense of hearing through time.

Sagan and his team raided the extensive ethnomusicological collections of the Smithsonian Institution to locate recordings that met the criteria, and it seems likely that one of those they found there was *The Land of the Morning Star*. In 1962, the anthropologist Sandra Le Brun Holmes and her husband, filmmaker Cecil Holmes, toured Methodist missions in the Top End, the most northern region of the Northern Territory, to make the film *Faces in the Sun*. At the mission on Milingimbi Island in Arnhem Land, she recalls in her autobiography, people would come to visit her after the day's work was over:

> During such evenings ... I recorded a number of
> beautiful songs, didjeridu solos and stories from the
> men. One man named Mudpo was a virtuoso on the
> didjeridu, able to make the sounds of birds at the
> same time as the wonderful resonant music rolled
> on uninterrupted. There were fast songs and slow,
> ghostly music about morkois (ghosts). These men

were masters of the instrument. It was the best music
I had ever heard, in the true classical, ceremonial
tradition.

Other participants, Djawa and Waliparu, are named in the
sleeve notes of the record Le Brun Holmes later released,
and it is their music that Sagan's team chose for the Golden
Records. Who were these master musicians and custodians
of their culture?

Djawa is well known: he was a community leader
and artist, and a winner of the 1955 Leroy-Alcorso Textile
Design Competition. The National Museum of Australia
has many of his bark paintings in its collection. Mudpo and
Waliparu are absent from the easily accessible archives, such
as those at the Australian Institute of Aboriginal and Torres
Strait Islander Studies (AIATSIS) in Canberra where I went
to seek more information. There is little trace of them apart
from the record's sleeve notes. To find out more, we would
have to dive deeper into mission records and talk to their
families and people who knew them.

According to *Murmurs of Earth* and the official tracklist
at the JPL, the one minute and twenty-six seconds on the
Golden Records included 'Morning Star' and 'Devil Bird'.
However, when the Golden Record is compared with the
original recording, it becomes clear that the didjeridu and
clapsticks (Mudpo and Djawa) is the first twenty-three seconds, with Djawa's vocal cut off, while the remainder is not
the 'Devil Bird' song at all, but Waliparu singing 'Moikoi'.

'Morning Star' is a clan song (manikay) relating to the

Barnumbirr morning star ceremonies of Yolngu people; such songs were like title deeds, expressing the relationship of families or clans to areas of land through the ancestral spirits. 'Moikoi' is about malicious spirits who try to entice newly deceased souls away from their clan country. The songs, in their new context on the spacecraft, could perhaps be read as a message about the journey of the human spirit between Earth and space – and home at last.

Working frantically to meet the launch date, there was little time for the Golden Records team to seek copyright clearances beyond the record companies who issued the music; and issues around Indigenous intellectual property weren't as well understood in those days. Sagan released a CD of the Golden Records for a new edition of *Murmurs of Earth* in 1992; but the names of the Indigenous artists were still missing. Djawa's, Waliparu's and Mudpo's names do appear on the track listing of a more recent release for the 40th anniversary of the Voyagers' launch, although it's not clear if their families were contacted.

This is how Sagan summed up the purpose of the Golden Records in 1978, the year following their launch:

> Our concern with time and our sense of the Voyager
> message as a time capsule is expressed in many
> places on the record – greetings in Sumerian, Hittite
> and !Kung, photographs of Kalahari Bushmen,
> music from New Guinea and from the Australian
> Aborigines, and the inclusion of the composition
> 'Flowing Streams', whose original structure

antedates Pythagoras and perhaps goes back to the
time of Homer.

Interestingly, the Indigenous groups mentioned here are
among those most often singled out in early anthropology
and popular conceptions as the most 'primitive' on Earth.
They are mentioned in the same breath with long-dead cultures known mainly from archaeology.

But these were not dead or dying cultures. Throughout
the 1950s, 60s and 70s, Yolngu people in Arnhem Land were
fighting battles to maintain their land and culture against the
erosions of missions, mining and exploitative art dealers. In
1962, when the recording was made, Aboriginal people were
subjected to the Australian government's pernicious assimilation policy which resulted in the Stolen Generations, and
denied them just wages for their labour. In the 1970s, the
decade of the Voyager missions, assimilation was superseded
by self-determination; yet the rights of Yolngu people were
easily discarded when mining interests were at stake. Now,
in the 2000s and 2010s, the battle goes on under the policy
known as the Northern Territory Intervention.

Le Brun Holmes does not mention the Voyager missions
in her autobiography. In 1977, she was busy campaigning for
Davis Daniels, an Aboriginal man from Roper River who
was standing for election in the Northern Territory. Perhaps
Djawa, Mudpo and Waliparu never knew that their music
had swept past Jupiter, Saturn, Uranus and Neptune towards
interstellar space on Voyager 1 and 2.

But in contrast to Sagan's well-meaning conception, this

music is not the preservation in copper of a vanishing way of life. It could also be read as a mark of the resilience and adaptability of Aboriginal culture, as it sails out of the solar system, far, far beyond the morning star.

CHAPTER 7

WHOSE SPACE
IS IT ANYWAY?

08 morning star song, Venus rising   comet dust string
to a lorikeet dawn, ironwood fire cracking, reverberation
     of the verse
stringybark   sugarbag   lines of song

Meredi Ortega, 'Liner Notes, Voyager Golden Record'

Someday we will live among the stars. There will be settlements on the Moon, cities on Mars, holiday resorts on Titan,
orbital habitats in the Rings of Saturn. People don't necessarily dream of being heroic astronauts risking their lives with
only the thin walls of a spacecraft between them and certain death. They want to experience the landscapes of other
worlds, as painted by the deep space probes, with their own
senses. Watching Earth rise over the Martian desert, skiing
on the snows of the moon Europa, or swimming like a fish

in the freedom of low gravity would all be experiences worth making sacrifices for; and if you can do this from the comfort of a space hotel, all the better.

Holidays are one thing, but for many advocates, space exploration is the answer to human woes. As we exhaust terrestrial resources, and stare down the barrel of rising sea levels, moving off-planet holds as many promises as it does challenges. We imagine these star colonies as utopias, where we leave behind Earthly conflicts and inequalities. Part of this is premised on the idea that our arrival will displace no Indigenous inhabitants and wreck no ecosystems. As far as we know, we are alone in the solar system. The resources of other worlds are there for the taking because no-one else is using them. This will be a pure form of colonisation where nobody loses.

The move beyond Earth will, however, depend on technology, as the human body is adapted to a very narrow band of gravity and atmospheric pressure. Human ingenuity will have to replace what Earth formerly provided: water, air and gravity. If we solve those technical issues, then there will be nothing to stop us.

But it also depends on who 'we' are.

THE 'SWEET POISON OF THE FALSE INFINITE'

Panspermia is the theory that the universe is filled with life. Just not the kind of alien life you might imagine. According to this theory, micro-organisms and prebiotic molecules –

these are complex compounds like amino acids that aren't alive, but occur in living things – travel on comets and asteroids between the worlds, flourishing when and where conditions are right. The theory proposes that expansion of life into every available niche is a natural process that's taken place countless times in this, and other, galaxies. Recently the strange elongated rock 'Oumuamua, from somewhere in the Milky Way, swung in and out of the solar system, giving astronomers a brief chance to survey it. This is precisely the kind of object that pansperm-ists hypothesise transported the seeds of life between star systems. To date, evidence that micro-organisms can sur-vive journeys in space, even if encased in meteoroids, is scant. While there's plenty of evidence of prebiotic com-pounds in our solar system, as Philae found on Comet 67P Churyumov-Gerasimenko, and in meteorites which fall to Earth, it's not clear how these might transition into unicel-lular or more complex living things. Critics also point out that the theory merely delays the real question of how life started.

While panspermia is controversial, the corollary of the theory is that enabling the spread of human life throughout the universe is justified, even a moral imperative. This view has been espoused by influential proponents of space explo-ration. Consider this from American science fiction writer Ray Bradbury in 1973:

What's the use of looking at Mars through a telescope, sitting on panels, writing books, if it isn't to guarantee, not just the survival of mankind, but mankind [sic] surviving forever!

This is space-travel advocate Marshall Savage from his 1992 book *The Millennial Project: Colonizing the Galaxy in Eight Easy Steps*:

> We need to rupture the barriers that confine us to the land mass of a single planet. By breaking out, we can assure our survival and the continuation of Life.

If you're tempted to think these views are a little out-dated, here's a more recent example. In June 2018, Elon Musk said, on Twitter:

> It is unknown whether we are the only civilization currently alive in the observable universe, but any chance that we are is added impetus for extending life beyond Earth. This is why we must preserve the light of consciousness by becoming a spacefaring civilization & extending life to other planets.

This isn't even about humans any more – it's Consciousness and Life, as some kind of semi-mystical force. When did humans get to stand in for all life? The idea that humanity has a moral right to propagate indefinitely, in whatever form that might be, is clearly still alive and kicking.

This view pre-dated the Space Age, and came in for some trenchant criticism from CS Lewis. Earlier I described how his astronaut character Ransom was kidnapped, and Ransom's response to being in space. In 1943, before V2 rockets started bombarding London, Lewis wrote a sequel to *Out of the Silent Planet*. In *Voyage to Venus* (also known as *Perelandra*), Ransom is pitted against the evil scientist Professor Weston, who's very much in favour of 'man' spreading to fill the universe and expresses this in similar language to Musk.

In the passage below, Lewis captures the tension between what we might now call an ecological position, and a colonialist one, in a way that prefigures contemporary debates:

> It is the idea that humanity, having now sufficiently corrupted the planet where it arose, must at all costs strive to seed itself over a larger area: that the vast astronomical distances which are God's quarantine regulations, must somehow be overcome. This for a start. But beyond this planet lies the sweet poison of the false infinite – the wild dream that planet after planet, system after system, in the end galaxy after galaxy, can be forced to sustain, everywhere and for ever, the sort of life which is contained in the loins of our own species – a dream begotten by the hatred of death upon the fear of true immortality, fondled in secret by thousands of ignorant men and hundreds who are not ignorant.

I suspect Lewis means moral rather than environmental corruption, but the contrast still works for the 21st century.

Lewis's phrase 'the fear of true immortality' may be read as an allusion to the eternity of a Christian notion of heaven; but he is right to point out that this type of thinking is rooted in a deep fear of mortality which makes the 'sweet poison of the false infinite' profoundly seductive. Lewis argues that a virtually endless universe is not an invitation to expand, in our own messy organic big bang, to fill all available niches; nor is it a palliative for the fear of death. For that, one must look inwards.

The fear of mortality is a usually unacknowledged aspect of what many in the space world assume is a universal human urge to explore and to propagate. Trophies – like Golden Records, red sports cars and national flags, are left in space as monuments to preserve memory and ensure continuity beyond the grave. They're tangible reminders of who was there: an object standing in for an individual, a nation or an ideology.

EXTERIORES SPATIUM NULLIUS

One of the most iconic images of the Apollo moon landings shows astronauts standing next to the US flag. Apollo 11 was presented and interpreted as an action taken for all humankind, but the 'planting' of the flag is a classic symbol of national colonisation. In Australia, there are numerous paintings depicting the moment when Lieutenant James Cook claimed New South Wales for England by planting the Union Jack flag, on the basis that Australia was *terra nullius*, the land of no-one. This is also the moment of naming, and the names create a relationship between the place of origin – in this case South Wales – and the new landscape.

The use of a flag in this way is a powerful symbol of claiming sovereignty over territory. It can also be used in more benign situations, such as the national flags planted by Mt Everest climbers to indicate their success in reaching the summit, and at Earth's poles. However, in the lead-up to the Apollo 11 mission, there was domestic and international debate about creating the impression that the US was making a territorial claim in violation of the Outer Space Treaty, which they had signed only two years before in 1967. NASA's Committee on Symbolic Activities for the First Lunar Landing considered using a United Nations flag, but rejected it in favour of the national flag. Congress altered NASA's funding conditions to prevent the use of flags of other nations, or international associations, on future expeditions financed solely by the US. It was touch and go; there might have been international protests, whether formal diplomatic ones or popular ones, but they didn't happen. People were so awed by what had been accomplished that the international community chose to interpret the national symbol of the US flag as 'an historic forward step for all mankind [sic] that has been accomplished by the United States', as the committee phrased it. A few expressed regret that a UN flag had not also been used. The compromise was to fix a plaque to the landing module with the words, 'Here men from the planet Earth first set foot upon the moon July 1969, A.D. We came in peace for all mankind [sic]'.

But the flag is still a troubling symbol. In the new race to get back to the Moon, its interpretation could be open to change. The Outer Space Treaty forbids territorial claims, but many, including space lawyers, national governments and entrepreneurs, think the principle that space is the common heritage of humanity is holding back the development of private space enterprise. When the film *First Man*, about astronaut Neil Armstrong's personal life around the time of the Apollo 11 mission, was released in 2018, many Americans were very upset that the actual moment in which the flag was placed in lunar soil was not portrayed. This was often accompanied by an insistence that the Apollo 11 mission was 100 per cent American. A new conspiracy theory seemed to be growing up around the flag: that the omission was evidence of a 'globalist' cabal to undermine the US and deny its rights to the Moon, even if only symbolically. (I'm very curious about how much overlap there is between the lunar conspiracy communities on this one.) I was reminded of a slogan I saw in a US magazine years ago, illustrated in the style we'd now call a meme. It said, on the background of stars and stripes, 'The Moon is ours. Don't be landing your stanky rocket on the Moon'.

Space is often seen as the very last frontier, ripe for conquest by daring adventurers. In this view, the trajectory of human evolution is a continual expansion into new territories, from the first steps our hominin ancestors took outside the African continent in a sort of Palaeolithic *lebensraum*, to the 'high frontier' of space.

Frontiers are associated with exploration and struggles

against hostile nature which are overcome by curiosity, technology and courage. On one side of the frontier is tame 'culture'; on the other wild 'nature'. Humans proved tremendously successful at adapting to new continents and environments using technologies such as fire, stone tools and metallurgy. By the 20th century, technology had enabled humans to move beyond the narrow band of pressure and temperature where our bodies had evolved, to explore the deep sea, Earth's poles and outer space.

The urge to explore, or natural human curiosity, is often called upon as the reason humans left Africa and why they must go to space. This urge is framed as an innate human characteristic that cannot be suppressed or denied. The idea that there are deep evolutionary roots that impel humans towards space is captured in a famous scene known as the 'Dawn of Man' [sic] from Stanley Kubrick and Arthur C Clarke's 1968 film *2001: A Space Odyssey*. Inspired by a weird alien monolith (we'll just set that aside for now), one of the hominin ancestors discovers how to use a bone as a weapon. In a conflict over access to a waterhole, the ancestor and his brothers (I'm sorry but they have fur penises) beat an individual from a rival group to death with bone clubs. It's the primal horde straight out of Freud's *Totem and Taboo*. The aggressor then flings the bone into the sky, where the scene cuts straight into a spacecraft sailing through Earth orbit to the Blue Danube waltz. Space, the scene seems to imply, is practically in our DNA.

This is a narrative that I hear repeated over and over. You'll find it in countless books, movies, academic articles, conversations and in social media when rationales for going into space are being discussed. My space colleagues frequently accept it without question. But as an archaeologist, I'm naturally sceptical of claims to a universal human nature. Curiosity is not a characteristic researchers generally include in behavioural modernity, that suite of behaviours that set human ancestors on our current path, and it's hard to know what kind of archaeological evidence would be needed to identify it. The almost universal cultural beliefs about the Moon and stars do not mean that people were nurturing a desire to go there. For much of human history space was not even seen as 'outer'; the idea of an infinite, empty geometric space that could be traversed or filled emerged after the 16th century.

I call this package of beliefs about space the 'Space Race Model', by which I mean a theoretical structure used to interpret the history of space exploration. Because it's presented as a natural inclination, the 'urge to explore' conceals the other motives for space exploration, such as military advantage, national prestige and access to resources. For something meant to be a universal characteristic, the urge is very unevenly distributed. The division of nation-states into 'spacefaring' and 'non-spacefaring' surely belies the idea that space exploration is a universal human endeavour arising from a deep evolutionary past. If human ancestors experienced the urge to explore, then contemporary Indigenous people, for example, are not assumed to have inherited it; they, and other groups, are meant to be the recipients of space benefits rather than the suppliers. I don't think you would ever hear anyone articulate it like this – they'd probably be horrified to have such views attributed to them – but it's implicit in the discourse. Who counts as human for the purposes of the 'human urge to explore' is in fact a narrow group of people.

Curiosity is not a straightforward quality either. For chunks of European history, it was considered a feminine failing, to be discouraged. Tales like 'Bluebeard' warned women of the consequences of being too curious; if you don't remember this sinister fairy tale, the consequences include a bloody death. High-profile examples like Valentina Teresh-kova aside, the first woman in space in 1963, the deliberate exclusion of women from astronaut training in the US until the 1980s and continuing resistance in Russia (a Russian astronaut trainer said to me 'Space is no place for a woman') shows that 'human' meant 'male'; women weren't supposed to have such urges! Heaven only knows how planetary colonisation was going to work without them. (Economist George Loewenstein conducted a review of the meanings of curiosity in 1994; I do rather like his argument that one of the salient characteristics of curiosity is a 'tendency to disappoint when satisfied'.)

In the Space Race Model, the interests of mostly white male American astronauts, space administrators, scientists and politicians are presented as universal human values. It's a particular (masculine) version of behavioural modernity that is co-opted to justify the current models of space occupation. In other words, all the trappings and assumptions of Western colonisation that have played out over centuries.

This narrative conveniently forgets the entanglement of

space with the Second World War, the Cold War and the Vietnam War, the slave labour which produced the original V2 rockets in Germany, and the nuclear weapons which were the rationale for continuing to build them. Tellingly, the Dawn of Man scene from *2001* very explicitly places a violent act at the heart of technology and culture in its portrayal of how the Space Age is implicit in evolution.

WHO HAS THE RIGHTS TO SPACE?

The Outer Space Treaty, which came into effect in 1967, was negotiated at the height of the Cold War, when nations feared one political bloc would gain control and shut everyone else out. The treaty proclaims that space is the common heritage of humanity, that space should be used for peaceful purposes, and that no-one can make a territorial claim. These are the exact words:

> Outer space, including the moon and other celestial
> bodies, is not subject to national appropriation by claim
> of sovereignty, by means of use or occupation, or by
> any other means.

The treaty known as the Moon Agreement, created by the UN in 1979, stipulates that the Moon and other bodies in the solar system should be used for the benefit of all people and all nations. The agreement promotes international sharing of resources and scientific samples. It also bans the use of the Moon for military bases or weapons testing. Few of the major spacefaring nations have ratified the treaty, for a very simple reason: they don't want to be left out if it does become possible to claim territory or resources on the Moon.

In all of this, the intrinsic values of the space environment have been frequently overlooked. While the need for an environmental ethics of space has long been recognised, there is little evidence that space industry has moved beyond a purely anthropocentric perspective: space is considered only in terms of what it can offer humans. On Earth, the concept of 'nature' having value in its own right, independent of human use, is no longer problematic. Australian philosopher Val Plumwood was at the forefront of a movement to break down moral distinctions between humans and nature. Plumwood argued that nature has its own agency or autonomy, and should be reconceived as a co-participant in human endeavour rather than something on which we are dependent.

In her view, we pay attention to the resources afforded by the environment, and the limits they impose on our activities, only when our access to them is jeopardised. The amount of space junk and the risk of collisions is one such threat to accessing the resources of space. Even so, the problem is still framed from an anthropocentric and geocentric perspective: in other words, how will it affect Earth? The value of this apparently empty space – and the new cultural landscape it now constitutes – is conceived entirely in terms of human use. Could we argue that space has intrinsic value, and as such is a place towards which we have a moral obligation?

The issue is perhaps clearer when we consider other planets. The view that inanimate celestial bodies have a right to exist undisturbed has been called 'cosmic preservationism'. One of the arguments is that the uniqueness of these planetary landscapes creates intrinsic value. There is no doubt – as human space exploration has repeatedly proven – that each object in space has its own story to tell. However, critics of cosmic preservationism claim it leads to the absurd position of rocks on Mars having rights.

In order to continue as a spacefaring species, and even perhaps to continue to live on Earth, we have to find sustainable ways to use the resources of space to survive. Our very presence on other celestial bodies, whether in human form or through robot avatars, changes them. They are altered physically and conceptually, becoming part of a human cultural landscape in a new way. We cannot land, sample, build settlements or mines and then whisk away as if nothing happened – our chemical and mechanical traces are now part of the planet, asteroid or moon. At this stage of human space exploration such impacts are minimal, and no doubt acceptable to populations on Earth. But this won't always be the case.

As well as changing our perspective to see space as a cultural landscape, another shift in perspective is to see space as place. Part of how we do that is through giving names to features and objects. It's not just about mapping and scientific nomenclature: the names are bridges to places we can never go, but for which we have mental visions and metaphorical associations.

A PLANET BY ANY OTHER NAME

Naming is a powerful thing. It reflects values and aspirations and tells the history of how we have come to know the solar system. As historian Paul Carter says in his book *The Road to Botany Bay*, spatial history begins:

> … not in a particular year, not in a particular place, but in the act of naming. For by the act of place-naming, space is transformed symbolically into a place, that is, a place with history.

> And, by the same token, the namer inscribes [their] passage permanently on the world, making a metaphorical word-place which others may one day inhabit …

Celestial geography mirrors the power relations of terrestrial politics. The long tradition of giving planets, asteroids and galaxies classical and European names reflected a world view which privileged the Western over Indigenous, Eastern and global south cultures.

How do places in the solar system get named? We already use the names of gods and goddesses given by the Romans, over 2000 years ago, to the most visible planets:

Mercury, Venus, Mars, Jupiter and Saturn. The existence of Pluto, Uranus and Neptune wasn't known until much later; but they were similarly named after classical deities.

Planetary geography really kicked off with the invention of telescopes in the 17th century. Celestial places were being mapped by European astronomers at the same time as places on Earth, in the era of European colonial expansion.

The lunar *maria* ('seas'), mountains and craters familiar to us today were mapped by the first real selenographer (charter of the Moon), the Dutch astronomer Michael van Langren, in 1645. (Incidentally, he also made the first known statistical graph.) He chose mostly European royalty and the notable scientists who were his contemporaries to name these features. These included a large crater named after the French queen Anne of Austria, now more famous as a character in Alexandre Dumas's novel *The Three Musketeers*, and a small crater after the Jesuit mathematician Jean Leurechon, who, among other achievements, wrote the earliest known description of how the ear trumpet works.

Twenty years later in 1665, Giovanni Cassini observed Jupiter's giant red storm. He called it, rather prosaically, 'Permanent Spot', to distinguish it from the shadows cast by the orbiting moons on the surface. (Yet more kinds of space shadows, if you needed some extra examples.)

Some of the earliest names on Mars were given to light and dark markings (albedo features) by English astronomer Richard Proctor in 1867. The names he chose were those of astronomers involved in Mars observation, such as the Reverend William Dawes, on whose map he based his own. He was a bit over-enthusiastic, though, and re-used the same names in different features – hence the Reverend Dawes was immortalised not once but six times, as an ocean, continent, sea, strait, island and bay.

By the 20th century this messy and ad hoc system of naming celestial places was becoming a problem for astronomers. When the International Astronomical Union (IAU) was established in 1919, they set up a committee to sort out all these names, and started a Gazetteer of Planetary Nomenclature, in collaboration with the United States Geological Survey. Since the 1950s, spacecraft have replaced the telescope as the principal means of solar system exploration; and the number of identified places and features has increased dramatically with flyby, orbiting, and surface missions to most planets and moons. The IAU has had to step up its activities because they all need names.

The original rationale was to systematise naming for scientific purposes, so that astronomers and planetary scientists could be sure they were talking about the same feature. However, these days the IAU also aims to represent diverse cultures in the naming process, affirming the principle that space is the common heritage of humanity. This can be particularly meaningful in the case of Indigenous cultures who have endured the ravages of European colonialism.

The IAU's principles, summarised, state that names should be clear, simple and unambiguous to facilitate scientific communication, avoid duplication (much like racehorse names), avoid political, military or religious significance, and promote diversity. Each planet has a theme relating to the Roman or Greek god after which it is named. Within each theme, there is plenty of scope for broadening the representation of terrestrial cultures, as happened with two recent naming campaigns.

REFLECTING EARTH IN SPACE

In 2011, the Messenger spacecraft slipped into orbit around Mercury, the planet closest to the Sun. The spacecraft was a collaboration between NASA's Jet Propulsion Laboratory (JPL), the Carnegie Institute for Science and the Johns Hopkins Applied Physics Laboratory.

Since Mariner 10's flyby in 1973, Mercury's theme has been artists of all kinds. According to the IAU rules, features are named in honour of people who have made outstanding or fundamental contributions to the arts and humanities (visual artists, writers, poets, dancers, architects, musicians, composers, and so on). There you'll find the Equiano crater, after Beninese writer, abolition campaigner and former slave Olaudah Equiano; and the Sei Shōnagon crater, honouring the 10th-century Japanese courtier who pioneered the list as a literary form. I was particularly pleased to find her there, as she has been a favourite writer since my high-school days. As a teenage girl, I probably thought *The Pillow Book of Sei Shōnagon* was some saucy tale, only to find myself enchanted with her wry and poetic vision of small, everyday things.

In April 2015, Messenger ran out of fuel and plunged to the surface, forming the planet's first archaeological site, a tiny crater splashed with the molten residues of metals and other human materials. I don't know if this crater has been named yet. Towards the mission's end, many features on Mercury were given names for the first time. The names for five craters were crowd-sourced: the final selections included Oum Kalthoum, the revered voice of the Arab-speaking world, and Enheduanna, a Sumerian princess from the 23rd century BCE who is the world's first known poet. As I found out later, the JPL staff were also given the opportunity to name some features, and one Australian scientist chose a legendary artist: Albert Namatjira.

Albert Namatjira (1902–1959) was an Arrernte man born in the Northern Territory of Australia. He was the first Aboriginal artist to become well known to the white Australian public, and his work won international acclaim. Growing up on the Hermannsburg Mission, he was initiated into Aboriginal cultural law at thirteen. In his twenties he learnt to paint from artists visiting the mission and quickly became proficient. His first solo exhibition was held in Melbourne in 1938, and a school of artists built up around him. His distinct style combined Aboriginal approaches to country with modern European landscape painting. His paintings represented the colours of the Australian landscape like no other.

He is best known for his evocative paintings of the MacDonnell Ranges. In his lifetime he received honours such as Queen Elizabeth II's Coronation Medal (1953) and Honorary Membership of the Royal Art Society of New South Wales (1955). In 1968, he was the first Aboriginal person on an Australian postage stamp. As an Aboriginal man of this era, he experienced the tensions between expressing his Aboriginal identity and placating Australia's administrative regime, which tightly controlled the freedoms of Aboriginal people. Nevertheless, he became one of Australia's most well known and popular artists with his work represented in major galleries. In 2002, the National Gallery of Australia mounted a major retrospective of his work, which highlighted how his art subverted the 'dead heart' of Australia as seen by European eyes, to show a vibrant and multifaceted landscape. This could equally apply to planetary landscapes like Mercury: the eye of the artist allows us to see beyond the superficial to the complexity within, as Alan Bean aimed to do for the Moon.

As a result of this, I decided to find out how well Australian Aboriginal culture was represented in the solar system by combing through the IAU Gazetteer of Planetary Nomenclature. It took me a while, as these things are never as straightforward as you think they'll be when you start.

Of over 15 000 officially named places in the solar system (excluding Earth), it turns out that approximately 0.3 per cent are Australian Aboriginal words. They are found on four planets, Venus, Mercury, Mars and Pluto; four moons of Saturn; one moon of Uranus; and four asteroids. Some places, particularly on Mars and the asteroids, are named after similar features or relevant towns in Australia, which just happen to have Aboriginal names. Most, however, belong to Aboriginal ancestral beings related to the naming theme of the planet. The theme for Venus is love; accordingly, there are eleven features named for Aboriginal female beings associated with fertility and creation. The names come from published collections of Aboriginal stories. One source was Katie Langloh Parker's *Aboriginal Legendary Tales*, first published in 1896. Langloh lived on 'Bangate' station in the north of NSW and she was told 'not only the legends, but the names, that I might manage to spell them so as to be understood when repeated', as she said. Her principal instructors were the Yularoi women Hippitha, Matah, Barahgurrie and Beemunny, who worked on the station.

While this research is very preliminary, the principal question raised is what the names mean in their new context. Across Australia, there are landscapes created by the same beings that are also featured in the Planetary Gazetteer: for example, the Wawilak (Wawalag) sisters on Venus. They are also the focus of major ceremonial cycles practised by contemporary communities. The use of some words is subject to restrictions based on gender, age, grade or moiety, while others refer to parts of the landscape for which there is specific knowledge or custodianship. These are not forgotten 'gods and goddesses', like many of the classical names so common in the solar system, but very potent symbols of continuing Aboriginal cultural practices against the formidable array of colonialist alienation technologies.

One set of names evokes a very particular landscape of northern Arnhem Land, in the country of Yolngu people, particularly around Milingimbi Island. These names refer to creation and death, ancestral beings, distant places like the island of the dead and the land of the morning star, and the Barnumbirr and the Djanggawul ceremonial cycles. Major versions of these ceremonies centre on Milingimbi Island, and it was also here that anthropologist Sandra Le Brun Holmes, in 1962, recorded the music that was later selected for the Voyager Golden Records.

The terrestrial landscape shaped by the Djanggawul is that lived by Yolngu people today, sustaining their law and culture. The colonialist structure of space exploration has translated these concepts and places into a new setting which spans the solar system. What it may mean to have key places and beings in a 'living' landscape extended to outer space is a question that only the communities concerned can answer.

These names can do more than just represent Aboriginal people in space: there is the potential to contribute to sustaining culture in the present. Many projects in Australia today are repatriating objects and knowledge, in the form of artefacts, photographs, diaries, stories, administrative records, names and more, to the communities from which they were taken. There are numerous successful language revival projects and a growing interest in Aboriginal astronomical knowledge.

A first step might be repatriation of the place names of the solar system. In the Djanggawul example, the IAU sources lead to detailed ethnographic accounts by anthropologists Ronald and Catherine Berndt, collected in the field from the 1950s to the 1970s. The communities, families, and perhaps some individuals with whom the Berndts worked, continue to live in the landscape of the Djanggawul.

The space treaties operate on a European concept of sovereignty and land tenure. Australian Aboriginal and other First Nations people around the world often have different law systems based around kinship, custodianship and responsibility, which are no less binding. The implications of translating landscape features related to this knowledge into other planetary locations, in terms of how that custodianship and responsibility are allocated, are potentially profound.

CONTESTED TERRITORIES

It's great that the IAU is attempting to be more representative of human societies in the way it assigns names to off-Earth places. On Earth, though, there are still many inequalities that remain to be redressed.

Rockets are dangerous things. They explode on the launch pad, explode mid-flight, and crash and explode in places they're not meant to. To test and launch them, ideally the launch path should not be over cities, towns or populated areas, but over the sea, or in places with low populations. For the densely populated European nations who acquired V2 rocket technology, this meant turning to the colonies: to places that were supposedly 'empty'.

When the British army officers flew over Woomera in the late 1940s, they saw an unpopulated desert just waiting for rockets. No-one asked what had caused the apparent absence of people. Since the 1800s, Aboriginal people had been alienated from their country by the usual array of colonial processes: massacres; removal into missions run by religious groups; theft of their land, and so on. Most

Australians considered Aboriginal people to be a 'dying race'.

In reality, the vast restricted area of the launch range overlapped with the Central Aborigines Reserve and the traditional country of several Traditional Owner groups and nations. A combination of drought, the expansion of missions and the dangers of the rocket range meant that many Aboriginal people were under pressure to leave their country.

The plans for the rocket range project did not escape notice, and there were concerns over the impact on Traditional Owners across the country. A protest movement uniting a broad range of groups and individuals was mobilised. Its leader was Dr Charles Duguid, a doctor and fierce campaigner for Aboriginal rights. Duguid was primarily concerned about the impact on the Pitjantjatjara people, due to his long association with the Ernabella Mission just outside the Central Aborigines Reserve, but like many others he was worried about the nuclear weapons and re-armament so soon after the end of the Second World War.

Duguid was backed by the Presbyterian Church and attracted many others who were equally horrified that the Australian government was intent on developing these new weapons. One of these was the athlete and Yorta Yorta activist Doug Nicholls (who was later knighted and became the Governor of South Australia). In 1936, Nicholls, together with William Cooper, Bill and Eric Onus, Margaret Tucker and others, founded the first all-Aboriginal political organisation, the Australian Aborigines' League. Nicholls campaigned against the rocket range in South Australia, and met with the Governor-General to express his concerns.

On 31 May 1947, the Presbyterian Church and the Woman's Christian Temperance Union packed the town hall in Melbourne for a massive protest meeting. At that meeting, the Rocket Range Protest Committee was formed with members from trade unions, women's groups, the Communist Party of Australia (CPA) and the Australian Aborigines' League. The Committee lobbied the federal government to respect the rights of Aboriginal people affected by the rocket range. Many of the protest's leaders wrote pamphlets, articulating their vigorous objections to the government's abandonment of any commitment to Aboriginal rights when the land was wanted to develop defence missiles for Britain. Nevertheless, the rocket range proceeded. Anthropologist AP Elkin, firmly in the camp that believed Aboriginal people were dying out, acknowledged that the range would have adverse impacts, but argued that it would just hasten a process already in train.

Although the desert was supposedly empty, the evidence of Aboriginal occupation was everywhere. A reconnaissance mission in 1947 found that:

> Lying on the ground, here, there and everywhere, were tens of thousands of artefacts, discarded knives and spear-points of flaked and chipped stone, dropped by unencumbered generations of black men, and one had to accept the artefacts at their face value. Where one found artefacts, one found water.

Water was a critical resource needed by the rocket range and the village, so Aboriginal land use was a factor in deciding locations. The stone tools, however, were considered to belong to the 'Stone Age' of the past, where Aboriginal people had been stranded in the white Australian imagination. They were not part of the 'Space Age' which was being built on top of the same landscape.

Nonetheless, the township and rocket range were given an Aboriginal identity through the process of naming. As well as the name Woomera itself, the streets of Woomera village, surveyed in 1947, were given Aboriginal names. The white Australians living in Woomera were very aware that their familiar names derived from Aboriginal language and embraced this. Residents of Woomera, like the colonial administrators of the preceding century, were fascinated by the culture they saw themselves as superseding. In the 1960s, a Natural History Society was formed at Woomera. They invited guest speakers such as Norman Tindale, one of the founders of archaeology in Australia, and undertook expeditions to the gibber desert to collect geological, biological and cultural objects, such as stone tools and artefacts like the woomera or spear-thrower after which the range was named.

The enterprising Jean Macaulay, wife of the Native Patrol Officer Robert Macaulay, set up a business from her home, selling Aboriginal art, and souvenirs like miniature weapon sets. It seems that they weren't sourced locally; the advertisements mentioned consignments of art from Arnhem Land. I have visions of French, German and Italian space scientists, out there for the ELDO program, browsing in the shop for artefacts to take home, perhaps never aware that Kokatha people were still living in the area. These were all distancing devices, which placed Aboriginal people in the past, with the archaeology, leaving the way open for the Space Age.

All the same, young 'half-caste' women were employed as domestic staff at Woomera, and Native Patrol Officers kept in touch with all those still living in the desert in order to warn them of impending launches. Some Aboriginal people resisted the pressure to 'come in' from the desert – they knew the implication was losing access to their lands – and were still living under the rocket path. A mural painted by an Aboriginal artist in 1980, on the walls of the old Heritage Centre, shows foraging parties watching the rockets flaring overhead.

Woomera is a very significant place in Australia's space history, and indeed, globally. If I were writing a statement of significance for a heritage study, I'd highlight the home-grown technology developed there, such as the sounding rockets and satellites. The theme of international cooperation in space is very strong, with collaborations with NASA and ELDO, the precursor of the European Space Agency. Until the 1970s, Woomera was part of all the major US space projects from Vanguard 1 to Apollo. All of this took place in a uniquely Australian landscape of gibber plains, salt lakes, red desert sands and mulga scrubs, inscribed with Aboriginal rock art and patterned with millennia of stone tools. The creation of the desert landscape was told in Aboriginal lore as the actions of the seven women who became the Pleiades constellation. In terms of historic, social, scientific and aesthetic significance, Woomera has it all.

Where it doesn't stack up is with the familiar narrative of the Space Race Model. Aboriginal people weren't thought to have an innate urge to explore, or a moral imperative to take their place in space. While some argued that the rocket range would provide employment for Kokatha and other people, this never happened. There was one Aboriginal fireman; and Aboriginal women were allowed in as domestics – this was the extent of the education provided to Aboriginal people at the time, preparing them for a life of menial servitude. The much-vaunted human characteristic of curiosity wasn't extended to include them. No Kokatha men were sent to the UK to get trained in aerospace engineering, as happened with a number of young white men. No Kokatha women were trained as 'computers', unlike the young white women who performed the calculations needed for the rocket launches in the era before electronic computers were common.

There's a painting in the South Australian Museum by Munggarawuy Yunupingu, one of the famous Yunupingu clan from Yolngu country in the Northern Territory. This is where a tracking station for the Europa rocket was located, on the Gove peninsula. In 1967 Yunupingu painted the Europa rocket filled with white people going up. Outside the rocket, black people are falling with their heads towards the Earth. It's a stunning representation of who was allowed to have curiosity and the technology to satisfy it. I go to look at it every time I'm in the museum.

To fully understand Woomera as a space place, we have

to put the rest of the story back in. The protest at Woomera had an important role in the development and growth of Aboriginal political activism. It was also a catalyst for debate about the participation of Aboriginal people in contemporary Australian society. The 1947 protests were only the beginning, as the Woomera region remains a favoured place for the Australian government to locate unpopular installations, such as the US military surveillance base Nurrungar, nuclear tests, nuclear waste dumps and detention centres for asylum seekers. As former Aboriginal Liaison Officer and Kokatha man Andrew Starkey described it to me, space didn't get a special pass because it was space. The Indigenous perspective understands space exploration as one strand of a colonial process that led to alienation from their country and deprivation of their human rights.

Woomera is just one space site where Indigenous people and the space industry intersect in the same landscape. Others are the rocket launch facilities at Colomb-Béchar and Hammaguir in Algeria, Kourou in French Guiana and White Sands in New Mexico. As the example of Woomera demonstrates, the development of space industry is embedded in colonial history and economic relationships. From a colonial perspective, both interplanetary space and the lands of 'primitive' people are *terra nullius*, empty wildernesses, or moral vacuums, into which 'civilised' nations could bring the 'right' moral order. The colonial aspects of space exploration are a mirror of those same aspirations played out on Earth.

In the discussions currently going on around the private use of space resources, some argue that corporations that put up the finances to develop and implement the technologies needed for space industry have a right to profit from the resources that no-one would be able to access if they weren't doing it. The treaties which insist on the common heritage of humanity and sharing space resources equitably are a disincentive for investors and entrepreneurs, they say. It's a fair point to make. But what about the contribution made by Indigenous people who were forced from their territory and marginalised in their own lands in order that wealthy, industrial Europeans could benefit from the land made 'empty' through genocide and dispossession? It was not a willing contribution, but in these places Indigenous people paid a cost for the 'curiosity' of others. When are they going to get a return on investment?

LINES ON A MAP

The Antarctic Treaty System is often used as a model for how space might be regulated. The Antarctic is defined as a global commons in the same way as space, and like space, there were no Indigenous inhabitants before European explorers went there. No-one can make sovereign claims, and military activities are prohibited. Nations are free to conduct science, sharing resources and results. So far the system has worked well enough. However, the US takes the position that space is not a global commons or the common heritage of humanity.

As a thought experiment, I sometimes imagine what

the solar system might look like if the Outer Space Treaty prohibition on territorial claims in space was watered down or removed. Space might be carved up between nations, space agencies and corporations, basically the people who have enough money to get off-Earth. In one scenario, the US might claim the Moon: it's very strategic as the gateway to the solar system, and as Albro T Gaul reminded us, 'Who rules the Moon controls the Earth.' How long would it take, though, before the Moon's water ice was depleted? Would the Moon be abandoned once its resources were mined out? And what kind of Moon would we see looking up into the sky at night?

Russia has the most experience in exploring Venus, but it's unknown if Venus has resources that could be used on Earth, and the cost of extraction in those harsh conditions may be too great. Mars might be divided up between the US, India, European nations, and private corporations like SpaceX. Private corporations will also lay claim to asteroids with their mineral resources. Other nations may have to settle for orbiting habitats, negotiating for access to ground resources with those who control the Moon or planet's surface. New kinds of solar system maps would be necessary, conceptual lines drawn over the deserts and mountains of Mars, asteroids depicted in the faded atlas pinks and greens to show which corporation has claimed them. I wonder how the sharing of scientific knowledge will play out when there is profit to be made from it?

It might be that nations and corporations will establish their claim to space territory by pointing to historic use – so the archaeological record which demonstrates that use may be critical. We've seen this happen on Earth, where Aboriginal people have had to establish their title to land in the courts by drawing on archaeological evidence to prove their connection to country. The heritage places where different nations and communities have left their traces in interplanetary space may become very important. And as we've also seen on Earth, heritage is political. Destroying or undermining someone else's heritage is a way to assert dominance.

As another thought experiment, what would happen if Aboriginal people laid claim to the planetary places now named for their cultural knowledge? Would a custodian of a particular place or piece of knowledge now have responsibilities for the interplanetary location, and would that constitute a sovereign right? I have no idea; but I think the question is worth asking. If space interests are going to overturn the idea of space as the common heritage of humanity, perhaps they should consider Indigenous systems of land tenure and management as a model. Flags don't have to define the frontier.

CHAPTER 8

FUTURE ARCHAEOLOGY

The wider our universe becomes due to science, and
the furthest we go – we think we go so far when we go
to the Moon – the nearer we need to come to the centre
of ourselves in order to interpret this world, in order to
find values, in order to give our lives meaning.

Anaïs Nin, Hampshire College, 1972

From time to time, I dream during sleep that I'm flying.
Standing on my feet, just a tiny movement will launch me
into the air, and I go soaring above the ground with my
arms outstretched, as effortlessly as a bird, delighting in the
freedom.

These dreams always have elements in common. The
weather is sunny with a blue, blue sky, maybe with a couple
of white fluffy clouds. There is green, even grass below me.
The ground is flat or gently undulating at most. There's

always a house. Not a familiar house, more like a greeting-card house or a children's book illustration. I fly above the house high, but not too high. I can see trees and often there's a Hills Hoist clothesline in the garden. It's much lower than aeroplane height, and the view is always of the immediate environment of the house. It's a domestic dream, and perhaps a child's dream, when your whole world revolves around the house and the extent of space is a concept you've yet to fully grasp.

When I wake up from a flying dream, I often still feel the lack of gravity. The sensation that I can just will myself into the air and fly stays with me until I put my feet on the ground and realise how heavy I am, and how adhesive gravity is. I feel intense sadness at this moment. After a few steps, the residues of the dream evaporate and I'm fully awake.

I'm dubious about the innate urge to explore, but humans do seem to have an urge to defy gravity. This starts very early. Remember how you loved being thrown up in the air and caught when you were a baby, and how it made you laugh? The thrilling sensation of a centrifuge, when an adult or older child held your hands and swung you around in a circle? And just how much fun swings in the local park were? Amusement parks offer adults the same excitement with their roller coasters, drop towers and large-scale swings. The playground rocket never lifted off, but if you climbed to the top, you could see the landscape spread out below you as in a bird's eye view.

My desire to fly doesn't mean I'm drawn to sky-diving, or hang gliding, because that's not what this is about. It's about flight being in one's body, not a result of technology. Having said that, I also have a slightly concerning desire to throw myself off heights just to experience the exhilaration of falling through the air – something called the High Place Phenomenon. Maybe this is like Douglas Adams' 'learning to throw yourself at the ground and miss'. Or, as my godson, Roy, says, 'the ground is so mainstream'. I'm not afraid of heights, just afraid that if there isn't a barrier, the compulsion might become too great to resist. My lack of fear frightens me because I know logically that hitting the hard ground at acceleration is not going to have a good outcome, but at some visceral, limbic level, I want to fall and be free.

For me, these dreams illustrate how gravity is something we take for granted until we are forced to think about it. Gravity can be understood through more than just an equation. It elicits emotional and physical responses which only become evident when we experience it in altered contexts. As I thought about the role of gravity in shaping human culture, I realised that the concept of the 'hyperobject' might be quite useful in understanding how we relate to it, and beyond that, to space itself. The philosopher Timothy Morton defines hyperobjects as 'entities of such vast temporal and spatial dimensions that they defeat traditional ideas about what a thing is in the first place'. I'm interested in how the barely imaginable dimensions of cosmic space and time manifest themselves in human bodies and thoughts. Terrestrial life is cocooned within a bubble defined by how far we can propel our bodies from the safety of Earth gravity without dying, but we can think and feel ourselves far beyond that.

TRUE INFINITE

Hyperobjects are not like other objects. You can't see them in their entirety, but you can perceive them through their effects. Once you have realised that a range of disparate effects are the manifestations of the same massive thing, it has moved from its unnoticed place in the background to the foreground of your consciousness. Then it filters everything about how you view the world. Or, as the common saying goes, once seen, it can't be unseen.

The solar system is something we can conceptualise, but not touch or see in its entirety. Usually it feels stable, like the mechanical solar system model called an orrery, where tiny artificial planets run on their preordained orbits. But when you look at the solar system's deep history, you realise that we're just living in a lull between catastrophic events like planets colliding and asteroid strikes. One such event is thought to have brought the age of the dinosaurs to an end, approximately 66 million years ago. Every year astronomers locate more and more objects in the near-Earth region, and the possibility of another major asteroid strike no longer seems like a random aberration but business as usual in the solar system. The next such event might not happen for a thousand years; but it could equally be next year. The chaotic, unstable solar system is a hyperobject that won't let us rest. The more you know about it, the more unsettling it becomes.

Hyperobjects are not located in a specific and precise place, or in a particular moment. They're huge and distributed over vast distances of space and time. A human might perceive a local instance or manifestation of a hyperobject, but this doesn't mean they can grasp the totality of it. Dark matter is a good example of this characteristic. We only know it's there because it seems to be missing in action. When all of the matter in the universe is taken into account, it turns out there's not enough of it to provide the gravitational force for galaxies to hold together in the way that we observe. There has to be some other mass out there. Dark matter permeates every corner of the universe but it's only known by this gravitational effect.

Laniakea, the supercluster of 100 000 galaxies that the Milky Way is part of, is another example of massive distribution through time and space. The cluster is 520 million light years in size, a distance that no human can ever travel. It's not a permanent structure and will drift apart at some unimaginable time in the future, just like terrestrial continents. If you look up into the night sky of the southern hemisphere and find the constellations Norma and Triangulum Australe, then you're looking through to the centre of the supercluster. Of course, you might have guessed it, the Milky Way is way out at the edges. Because of the data which was used to define Laniakea, pictures of it look like a curved and luxuriant ostrich feather: the galaxies in the supercluster are not represented as dots, but paths of movement, the direction they're travelling in. When, as a child, I looked up towards the band of the Milky Way, I was seeing a tendril of Laniakea with the full glorious structure lying light years behind it.

We could even argue that the cultural landscape created by space junk in Earth orbit is a hyperobject. It's spread over an area far vaster than Earth's. No one image can show us what it looks like. A photograph of a section of sky might show several satellites as points of light, but you can't see all 23 000 large objects at once, and you certainly can't see the minute particles. We have to estimate what is there by using multiple sources and predictions, and then use this data to make images and simulations which are more easily digested.

This is a very archaeological way of looking at things – studying the traces of a hyperobject in order to piece together the bigger picture, using multiple strands of evidence and techniques. The information that Voyager 1 and 2 gather in their instruments about the region of space they're currently passing through is similar. The data they send back to Earth are interpreted by scientists to make conclusions about the complex structure of interstellar space.

Space anthropologist Michael Oman-Reagan argues that interstellar space is itself a hyperobject. If you think of it this way, he says, we encounter it at many different scales. It's not something remote and desolate and impossibly far away: it's close and intimate as well. The signs by which we recognise it in everyday life don't require instruments to read. Oman-Reagan says we can perceive the traces of interstellar space in writings like those of Carl Sagan and Herman Melville – and in the feelings their words evoke.

I've tried to read *Moby-Dick*, Melville's classic novel about whaling in the 19th century, I really have; and I've never been able to get beyond the first few pages. But I did find out what Michael meant. The novel is about Captain Ahab's quest to kill the white whale which injured him. One day, the main character Ishmael looks over the edge of the boat to see the whale lurking under the surface of the ocean. He thinks to himself:

> Is it that by its indefiniteness it shadows forth the
> heartless voids and immensities of the universe,
> and thus stabs us from behind with the thought of
> annihilation when beholding the white depths of the
> milky way?

Interstellar space is hidden in the depths below the whale, whose milky white body only serves to accentuate the form-less space in the dark sea. It's an 'as above, so below' moment in reverse. The 'stabbing from behind' immediately made me recall a concept from Frank Herbert's *Dune* novels. This is *adab*: 'the demanding memory that comes upon you of itself'. At some deep level the void is always present just beyond the limits of memory and sense, until it forces itself into your consciousness. Plato had a similar concept in *anamnesis*, the theory that all knowledge was really just remembering things that we already knew from an eternal existence.

This isn't a knowledge which involves physics and mathematics; it's emotional and psychological, and no less valid for that. If I can make another connection here, then I'm reminded of Freud's theory of the uncanny, the thing that is horrifying because it's so familiar, like the doppelganger: the white whale doubling as the Milky Way and calling the void forth into one's soul.

It doesn't always have to be horrifying though. The white whale might summon forth the oceanic feeling rather than the heartless void. This is a feeling of oneness with the universe that French writer and philosopher Romain Rolland, drawing on Hindu and Sanskrit mysticism, thought was the origin of the world's religions. He wrote about this in a famous letter to Sigmund Freud in 1927; and the idea unsettled Freud so much that it took him two years to reply (a fact which makes me feel a lot better about my out-of-control email inbox). The comparison of space to the ocean has been made many times, by people such as Arthur C Clarke, Carl Sagan, and space historian William E Burrows. Contemplation of the ocean can be a door to a transcendent experience, to a cosmic ineffable that expands the soul.

This experience seems to be temporary and evanescent. You're not in that moment all the time. Hyperobjects might be demanding and 'viscous' in the way they shape your perceptions, but they're also forgettable. You glimpse them, and they can be overwhelming, but they're hard to hold in your mind. You need something to recall them and make them stick. Something may jolt you out of the forgetfulness, like the re-entry of a piece of space junk, or an asteroid nearing

Earth, or the plastic bag at the bottom of the Mariana Trench, but you gradually slip out of your new way of thinking and back into the old, familiar one.

Gravity is a hyperobject with a fluid relationship to memory in this sense. It fundamentally structures everything about life on Earth: the shape of bodies, blood circulation, movement, architecture, cooking. But you only know it by its effects, like falling and staying still. It's so ubiquitous as to be invisible and unremarked upon. Even when you fall over and scrape your knee, you curse your own clumsiness rather than gravity, which escapes blame. Going into orbit disrupts expectations and brings gravity dramatically into the foreground so you can't forget it. There's no up or down; objects float instead of falling; all your habits formed in gravity are suddenly irrelevant. Astronauts adapt pretty quickly though, and by the time they return to Earth, they've forgotten about gravity again and are surprised when things drop to the ground when let go (like toothbrushes and cups full of tea). Hyperobjects, it seems, have a tendency to form habits.

THE BODY IN THE MACHINE

It's one thing to encounter the void from the safety of Earth gravity, with the release of forgetting both the horror and the warm embrace of the oceanic feeling. But what happens when you encounter interstellar space at closer quarters? A chilling view of the psychological effects of space travel was imagined by a science fiction writer in the 1940s.

Cordwainer Smith was the pseudonym of Dr Paul Line-barger, a US expert in psychological warfare (or propaganda), and sometime CIA agent. The pseudonym was also used by his second wife, Dr Genevieve Collins Linebarger, a linguist and political scientist. Both she and his first wife, Margaret Snow, likely contributed more to his science fiction than is recognised.

Paul Linebarger's best known work is *Psychological Warfare*, first published in 1948. He and Genevieve were experts in Chinese and Japanese politics and international relations. In the 1950s and 60s, the Linebargers spent a lot of time in Australia as Paul was a visiting professor at the Australian National University in Canberra, in the Faculty of Modern History. Their experience there had a profound effect on their writing. In Cordwainer Smith's far future universe, Australians settled a planet called Old North Australia (Norstrilia for short) and still had the Queen of England as their head of state, 15000 years on, although no-one knew who or where she was. (Perhaps he would not have been astonished to learn that in the present all attempts to make Australia a republic have been defeated and the British monarch is still nominally the head of state.) Norstrilia is populated by austere and resilient people who make enormous wealth breeding sick sheep which produce an immortality drug called Stroon. It's the most valuable commodity in the universe. Even on another planet, Australians were still living off the sheep's back! The Linebargers planned to retire to Australia but sadly Paul died in 1966 before they could realise this dream.

A story of Cordwainer Smith's I find myself coming back to over and over again is 'Scanners Live in Vain' (1950). When it was written, it was not known whether humans could survive in space. Or even what space was truly like – the geomagnetic storms, plasmas, corrosive elements and radiation. All this would have to wait for the first satellites. In the early days of space, people sent all kinds of animals up to see if they would survive: cats, rats, mice, rabbits, spiders, dogs, monkeys, tortoises, cockroaches, and many other insects. It was a veritable Noah's Ark up there. I've even heard of an experiment where some human brain tissue was sent into orbit (although I can't verify it). It's easy to forget that when Yuri Gagarin made his first orbit in 1961, his survival was by no means a given.

Smith didn't imagine physical hazards for the first astronauts: rather, space was a psychological hazard. People in space did not die of radiation or exposure or extreme temperatures. They died of pain, the pain of space. Just to be in space was to endure unimaginable pain. To overcome this, spaceship crews were made into 'habermans', named after the inventor of the technology: their bodies were surgically severed from their brains and run by machines, so that they couldn't feel. As Martel, the protagonist in 'Scanners', says, '… you know what I am. A machine. A man turned into a machine. A man who has been killed and kept alive for duty.'

He says some revealing things about this condition to his wife, Luci, who is still fully human. He's just like the satellites whose death is a shadowy and ambiguous state of being:

Don't you think I remember what it is to be man and
not a haberman? To walk and feel my feet on the
ground? To feel a decent pain instead of watching
my body every minute to see if I'm alive? How will
I know when I'm dead? Did you ever think of that,
Luci? How will I know when I'm dead?

The pain of space is almost malevolent, like surf crashing
against a rocky outcrop, trying to get past the blocks to the
sentient being within. Any sensation would give it a tiny
crack to enter the mind and overwhelm it with pain: the
body is the route to being consumed by the void. The senses
have to be obliterated so the mind can withstand the pressure. Martel can't taste, feel, hear, touch. He monitors his
physical body with a control box set in his chest.

He's not a robot, but he's not quite human either; and as
the story unfolds, the ways of thinking that arise from being
cut off from the senses become quite important. But I won't
go any further in case you haven't read it. What I can tell you
is that in the story, the pain of space is overcome by lining a
spaceship with layers of life. Oysters were in the outer layer,
closest to space, and they died in agony. The animal and
human test subjects in the centre of the ship survived. Fortunately, this was the early test phase. Other means were found
to protect the human brain from pain, without the suffering
of shellfish.

The experience of pain seems very fitting for a psychological war expert to bring to an understanding of space.
What I find interesting about this is that it's a psychological

problem for which technological solutions (and a lot of oysters) are sought. The source of the pain is not explored, but it seems likely that it is caused by the feeling of annihilation that Ishmael the whaler experiences when looking at the shadow of the white whale under the water. It's no coincidence that psychologists played a very important role in early human space exploration – for example, think of Dr Bellows in *I Dream of Jeannie*. It was an acknowledgment that we just didn't know whether human sentience and consciousness were so earthbound that they could not survive beyond Earth. In those days, for all we knew, space was only ever going to be the province of robots.

SPACE MARKED BY DEATH

Martel asks his wife, Luci, a question I visited with zombie satellites in chapter 4: how will he know when he's dead? We can consider how to live in space, and the kinds of societies we might need, but we also have to think beyond living bodies. It's not something we've had to come to grips with so far. Space has barely yet been sullied by human death, partially because, with the exception of the Apollo astronauts, most humans in space stay within the relative safety of Low Earth Orbit.

There have certainly been space-related deaths. In 1967, three Apollo 1 astronauts burnt in a horrible pure-oxygen fire in their rocket, before they even launched, and cosmonaut Vladimir Komarov was killed on impact when the parachute of his Soyuz 1 re-entry capsule failed to open properly. The greatest loss of life occurred in the space shuttle Challenger (1986) and Columbia (2003) disasters, one on ascent and one on descent. No-one survived.

The only deaths that occurred while actually in space were three Soviet cosmonauts in 1971, when their Soyuz return vehicle from the space station Salyut 1 lost a valve and decompressed. When the recovery team located the capsule after its descent back to Earth, the cosmonauts were already dead.

The Starman, sitting silently in the Tesla Roadster, was never alive.

When Apollo 11 launched, the possibility that the astronauts might die on the Moon was a very real one. A speech was prepared for US president Richard Nixon to read if this occurred:

> Fate has ordained that the men who went to the moon
> to explore in peace will stay on the moon to rest in peace
> ... For every human being who looks up at the moon in
> the nights to come will know that there is some corner
> of another world that is forever mankind [sic].

The last sentence very clearly indicates the strong emotions which would result if the Moon became a cemetery – the ultimate Cold War memorial. I'm reminded of the First World War battle site of Gallipoli, where the burial of Australian soldiers on Turkish soil has given this place a special resonance in Australian heritage. There's a statement inscribed on the memorial to Turkish leader Kemal Atatürk in Canberra (although he did not actually say it) that always makes me cry:

> You, the mothers who sent their sons from faraway countries, wipe away your tears; your sons are now lying in our bosom and are in peace. After having lost their lives on this land they have become our sons as well.

Although no longer living, these bodies are active participants in making meaning at Gallipoli, even after more than a century has passed. Thankfully the Apollo astronauts all survived their sojourn in space, even the Apollo 13 crew.

There are some remains in space of people who died on Earth. A sample of astronomer Clyde Tombaugh's ashes was sent on the New Horizons mission to Pluto, the planet he discovered in 1930. Planetary scientist Eugene Shoemaker is the only person whose mortal remains are buried on another celestial body. Shoemaker was famous for his extensive study of impact craters on Earth, involvement in the Apollo missions, and co-discovering Comet Shoemaker-Levy, with Carolyn Shoemaker (he was her husband) and David Levy. He and Carolyn were driving to look for craters in the Northern Territory in 1997 when they were involved in a head-on collision, and Shoemaker was sadly killed. In 1998 planetary scientist Carolyn Porco designed a capsule to convey a portion of his ashes to the Moon, on the Lunar Prospector space probe. The probe was launched in 1998 and crashed into the shadows of the lunar pole in 1999. A quote from Shakespeare's *Romeo and Juliet* was inscribed on the brass foil wrapping:

> And, when he shall die
> Take him and cut him out in little stars
> And he will make the face of heaven so fine
> That all the world will be in love with night
> And pay no worship to the garish sun.

There are a number of companies who will launch a sample of your ashes inside a spacecraft into Low Earth Orbit. These all re-enter, however, for a second cremation in Earth's atmosphere. Among the people who've had their ashes in temporary orbit are *Star Trek* creator Gene Roddenberry, and LSD advocate and counter-culture guru Timothy Leary.

So there are a few cremated remains currently in space. There may also be some liquid and solid human waste from the space stations – the USSR space station Mir, for example, was said to be surrounded by a sparkling cloud of frozen urine. The human waste material is in such a low orbit that it's more likely to re-enter than stay up there. But thinking about this with an archaeologist's trowel in my hand, I observe there are no human bodies, body parts, or skeletons in space. Unlike Earth, which is layered with bones and teeth, the most durable parts of the human body, space is missing a key component of the archaeological record – at least for now. A future alien archaeologist surveying the archaeological record of the solar system would have a very mixed bag of evidence indeed from which to work out what kind of being made all the spacecraft and probes scattered so sparsely across interplanetary space.

But just think about the first human death while in space, or on a planet, rather than before or after orbiting. Think of space littered with dead bodies or lost astronauts. Imagine murder, massacre, genocide, warfare in space. Might this day come? Like the shells of dead spacecraft, these bodies will persist. There are no biological processes to break them down. There are no predators to gnaw on the bones. Cold temperatures will freeze or mummify them more efficiently than polar ice. On the Moon, ultraviolet and cosmic rays will slowly erode the spacesuit until the frozen flesh inside is exposed. On Mars, storms may hasten this process, stripping and burying the bodies under layers of red dust. In orbit, the human dead will not lie still, circulating among the dead satellites and the living.

I don't know exactly how the symbolism of death will change the meaning of space, but I know something will be different afterwards. If we come back to the idea that there exists an imperative to spread life throughout the universe, then we must accept that this also means spreading death.

WHEN LIFE MEANS GRAVITY

There'll be many new ways to die in space, and many new ways to live too. Hays and Lutes, in their theory of space power, proposed that wealth was the commodity driving space exploration in the current era. I've got my own theory about the basis of power in the next phase of a space-based economy: I propose that gravity will become a precious commodity.

Crew on the International Space Station have to exercise for over two hours each day to maintain their bone density and muscle mass. Living in gravity so different from Earth's has consequences, and without this regime, the astronauts and cosmonauts would end up with a condition well known to the post-menopausal women of Earth: osteoporosis, where the bones weaken and are prone to fracture. Effectively, if people want future settlements in space to succeed, there are going to have to be some longer term solutions to living in reduced gravity, and this will be critical for women's participation. Venus of Willendorf has to become an astronaut.

I investigated the social implications of this in an experiment with my students at the University of Applied Arts in Vienna, where I taught a course a couple of years ago. I drew inspiration from Stanislaw Lem's dark science fiction spoof *The Futurological Congress* (1971), in which the enigmatic Professor Trottelreiner proposes a method of exploring the future through language. The theoretical basis of futuro-linguistics, as explained by the professor, is this:

A man [sic] can only control what he comprehends, and comprehend only what he is able to put into words. The inexpressible is therefore unknowable. By examining future stages in the evolution of language we come to learn what discoveries, changes, and social revolutions the language will be capable of some day reflecting.

It's basically an exercise in wordplay: you take an existing word and add prefixes or suffixes, change letters, chop it up, make it into a verb, or whatever you please. Then you think about what the new word means and what kind of society needs this word to describe its values or practices. I was curious to see what happened when the class took 'gravity' as its base word. These are a few of the ideas we came up with.

A *monograv* is a person who has only ever lived in one type of gravity, considered inferior to those who are adapted to multiple gravity regimes. *Antineogravitationists* are people who oppose living in artificial gravity, believing that the body should adapt to whatever gravity it finds itself in. Gravity might become as contested as diet is now – perhaps there are groups who believe in *palaeogravity,* replicating the gravity conditions of the pre-spacefaring age. A *gravault* is a prison where people adapted to low or microgravity are incarcerated in higher gravity, their ability to move freely taken away. Perhaps this is punishment for the future crime of *kleptogravity*, where a person steals another's gravity ration.

Futurolinguistics is playful, but this made me think about how kleptogravity might work in a future space-based society. The earliest space stations, such as the rotating wheel conceived by the Viennese engineer Hermann Noordung in 1929, were designed to produce 'artificial' gravity by spinning, like the centrifuge astronauts train in.

Noordung thought that humans might be very uncomfortable with the constant sensation of falling in Earth orbit, but he reasoned that pilots become used to a variety of such experiences, so perhaps it wasn't impossible to adapt. In the 1960s, NASA considered spinning space stations, but decided that astronauts might not enjoy being spun like a sock in a washing machine. Having been in a gravity rotor in an amusement park, I can categorically say that I did not enjoy it.

None of the habitable space stations launched so far – the Salyut series, Skylab, Mir, Tiangong 1 and 2, and the International Space Station – have tried to create gravity by spinning. Part of their purpose is to conduct science in microgravity, after all; gravity is just a luxury for the comfort of the crew. The crew exercise regime is critical to ensuring that their bodies can adapt again to Earth gravity when they return. The treadmills, cycles and weights are a form of gravity surrogate, substituting for the work done by bodies going about their daily tasks within full Earth gravity. If you go to the gym on Earth, this is additional work; in space you exercise to replace such incidental activity.

Everyone gets equal access to the treadmills, weights and cycles. It doesn't take much to imagine, though, that situations might change and one group might assert dominance by controlling access to exercise, knowing the consequences for those excluded.

What if there is no return to Earth, whether that is for cultural, political or economic reasons? What about settlements on the Moon, with one-sixth Earth gravity, or Mars, which is about one-third, or permanent orbital habitats? Who holds power may be determined by who controls the access to specially designed high-gravity environments where people can maintain their strength. Gravity may indeed be rationed, and depriving people of access could have serious health (and even legal) consequences.

This may be particularly important for women in space. Gravity and gravid – a term meaning heavy with unborn offspring – both derive from the Latin *gravis*, meaning weighty or heavy. Evidence indicates that lack of gravity causes abnormalities in mammal embryos and foetuses, so pregnancy in low gravity is high risk. On the upside, one study suggests that microgravity might relieve some of the burdens on muscles and bones experienced in late pregnancy. Beyond pregnancy and birth, women may need different access to gravity than men to manage bone density loss before and after menopause. Controlling how women access higher gravity could become a new method of controlling female fertility and access to other resources; there's no reason to be optimistic about women's autonomy around this in space at present, when we watch what is happening on Earth.

Living in variable gravity environments makes gravity a commodity that it simply isn't on Earth; it becomes foreground rather than background, and there are all kinds of social implications to consider. They don't all have to be serious, though. One student proposed the *groovity club*, where dancers go to bust their moves to the DJ playing gravity levels along with the beats.

These might be futures where humans have spread

out into the solar system, to become a multigravity species. There are other futures too, where this doesn't happen. For an archaeologist, this is what one of them might look like.

THE ABANDONED SOLAR SYSTEM

It might be twenty years, fifty years or a hundred years in the future.

We're approaching the solar system from some other part of the Milky Way galaxy, to conduct an archaeological survey. I don't know who we are, or where we've come from. But let's assume that we have some kind of propulsion method that will allow us to travel rapidly, and sensors that allow us to detect anomalous materials and structures. We're looking for evidence of other sentient beings.

Our survey vessel is still coasting the interstellar winds when something unusual shows up, the first evidence that this system with its white star might produce results. At a bit over one light day from the sun, the sensors pick up a tiny body high in aluminium with long booms and a curved 3.7 metre dish. (This is Voyager 1.) The surfaces are pitted and dull and the spacecraft is silent, but for one thing – radioactive beeps emitted by twenty-four spheres of plutonium-238 oxide inside a cylinder delicately suspended from the body, and by uranium-238 coming from a tarnished disc attached to its flank. There are some scratchings on the disc but their meaning is opaque. We scan the entire spacecraft, and take some dust samples from it for later analysis. One thing puzzles us. It's moving away from the sun – a messenger, or fleeing what lies in front of us?

The survey vessel continues on its path. I might as well give it a name – from whatever original language we are using, the closest in English might be Insufficient Data For a Meaningful Answer, or IDF for short. When we're about six billion kilometres from this system's sun, we look in the direction the spacecraft came from and see a pale blue dot: one of the planets we'll survey soon and perhaps the origin of the dusty spacecraft.

Now we're inside the bubble of the heliosphere and sheltered from the interstellar winds, catching some solar rays. There are more comets and icy bodies (this is the region of the Kuiper Belt). The sensors pick up another spacecraft. Like the previous one, it's also travelling outwards, but the surfaces are much less weathered. It hasn't been in space for as long. It's not transmitting any signals but every now and then an alpha particle pops out of a cylinder of plutonium-238 that we surmise once provided power to the object. It's half the size of the previous spacecraft and we wonder if it was made by the same culture – the stylistic similarities, including the 2.1 metre dish, suggest yes, but this one encased in a thin golden film. (This spacecraft is New Horizons.)

A small planet with five moons (Pluto) looms into view. There are no biological or cultural traces on this dim red planet. But our scanners do detect a small wake of unusual molecules which suggest a spacecraft has passed by. This wake turns into a tenuous thread we find winding between the planets. The next planet is a blue ice giant with thin bright rings (Neptune). Its small rocky core is swathed in hydrogen, helium and methane gases. Fourteen moons orbit the planet; but there are no traces here. The next ice giant (Uranus) is similar but colder. Our survey covers the surface and the twenty-seven moons without finding anything.

The IDF continues towards a pale yellow gas giant with spectacular rings (Saturn). For a brief moment we wonder if they have been deliberately engineered, but they are natural – composed of dust, ice and rock. There are at least sixty-two moons. We start with the largest (Titan) and immediately hit paydirt. For the first time there is an artefact on a solid surface. It's a silver ribbed aluminium shell 2.7 metres in diameter, filled with metal parts but cold and dead, no working battery or radioactive fuel, and covered in layers of methane ice. The material is in reasonable condition as methane doesn't react with aluminium. This isn't the case for the remnants of what looks like a parachute used for braking, crumpled a short distance away. (This is the landing mission Huygens, deployed from the European Space Agency's Cassini orbiter, which has long since burnt up in the atmosphere of Saturn.)

So far we have two fleeing spacecraft and one lander. It's not much to go on but we hope to find more as we get closer to the central star. There's another gas giant ahead, but before we get there, the sensors react to something on a comet that's passing by. We draw in to have a look and find that this lumpy rock is covered in artefacts. And new technology: for the first time we see the long blue vanes of solar panels in a spacecraft which clearly crashed on the surface, splitting into pieces. Inside we can see the electronic entrails: a goldmine of technological information. On another part of the comet is a six-panelled box stuck under a ledge. Its intact state suggests that it successfully soft landed. Both objects are weathered from the jets of gas and dust that stream from the comet, but they seem to be of a similar age – perhaps different components of the same mission. (Indeed, this is Rosetta and Philae on comet 67P Churyumov-Gerasimenko.)

Our next port of call is the gas giant with swirling storm bands of orange and cream and a red spot (Jupiter). It has even more moons than the previous planet among its faint rings, including four very large ones, but none of them has traces of technology. A tiny shadow of metallic molecules may betray something which was once in orbit long ago but there is nothing there now.

After Jupiter the IDF hits the asteroid belt, a wide orbit filled with rocky, metallic and sometimes icy bodies, and a dwarf planet with bright spots (Ceres). Now we're really making progress as there are artefacts everywhere, although none of them are active. Orbiting the dwarf planet there's a little box with broad blue solar wings (Dawn). One asteroid has a box with shorter solar vanes on the surface (NEAR Shoemaker on Eros 433). Another is busy with stuff: six craft rest on its surface (Hayabusa2 with its rovers and landers on Ryugu). A little 10-centimetre blue cylinder (Minerva) is orbiting along with all the other asteroids. There are landers on two more asteroids (Psyche and Bennu). This is a veritable zoo of different robot species.

Before we leave this region we stumble across something truly different. The corroded, curved body has no obvious power source, or antennas, and has suffered much radiation damage – it was clearly launched without being strengthened against the harsh environment (this is the Tesla Roadster). From remnants of the original surface it seems it was once a vibrant red colour. The shape of what's left of it is nothing like that of the other craft we've seen so far. Compared to the other spacecraft it has a high carbon content and some molecules that we call prebiotic. Very mysterious. There are no other signs of life but we feel we're getting close to something now.

On the other side of the asteroid belt is the first planet with a solid surface (Mars). This dusty red planet is even denser with material, both orbiters and landers, of which there are at least fifteen. We're hopeful that this might be the origin of at least some of the spacecraft the survey has revealed so far but a full surveillance of the planet shows no signs of industry, habitation or organic life. Apart from the sixteen orbiters circling the planet and the dust devils whirling furiously, nothing moves on the surface. The sensors detect some landers buried in deep sand drifts. There's a new kind of artefact too, and when we find a small trace of tracks that have survived many storms, we realise that these are rovers designed to travel across the terrain. There are six of them, stopped in their tracks, but this is the first evidence of non-natural movement we've seen. The red dust has abraded and infiltrated every surface so it's hard to work out what the relationship is between the three types of craft.

After recording everything by remote sensing, we head for the next planet and its one large moon.

The pitted grey body of this moon has virtually no atmosphere and everything is laid out on the surface for us to record. There are around thirty orbiters of many configurations. There over sixty surface locations, and some of these are not just a single object as we've seen on all the other bodies so far. There are artefacts clustered around landers and in some locations, more than one lander. The rover types from the red planet are here too, but they're very, very different and in fact look a bit more like the strange vehicle from the asteroid belt. The empty rovers provide clues about the shape of the body which used them – the cavities, which those on the previous planet lacked, allow dimensions and limbs to be postulated. The flat-topped types are represented here too. The tracks that were erased in the red planet's winds are preserved clearly on this surface.

And there's something else: small, slightly elongated, striped indentations in the soft regolith. Hundreds of them, but only associated with one site type. Hypothesising that they may be the trace fossil of an organic life form, we estimate the height and weight of the bodies. It seems to align with the size of the rovers, so perhaps there was just one species present. At each of the seven locations these prints occur, it is possible to distinguish two individuals from the footprints and the cavities in the rovers suggest each visit was undertaken by a pair. Small differences in the size of the prints suggest a minimum of six individuals. Of course it's pure speculation without more evidence, but perhaps they are breeding pairs, or genetically related in some social unit.

At last we are getting closer to understanding at least one of the species that may be responsible for the technology we've observed. This is exciting, and already there are years of work ahead to analyse the data. The theory we're working on now is that either the planet this moon orbits or the next one will be the home of at least one spacefaring species. Given, however, that we have not found one active spacecraft yet, it's unclear what we will find. It's almost as if this solar system has been abandoned.

Turning our attention to the blue planet, many things begin to make sense.

At first we thought this planet had a ring, like the outer planets. Its spectral signature, however, shows it is composed mainly of metals and carbon rather than rocks and ice. And inside the ring, closer to the planet's surface, is a spherical cloud of similar debris. It is clearly not natural. The objects are highly fragmented from constantly colliding with each other, as if they had been smashed through a sieve. There is a miasma of dust enveloping the shards. The cloud is a brilliant diamond aura, catching the light of sun, moon and planetshine as it tumbles and swirls. The IDF is well shielded, but it's not worth taking the risk of approaching this entombed planet much closer.

In the outer ring, the satellites are more intact and there is some chattering and radio noise, coming in fits and sputters. We catch one for later study. It resembles the craft we'd seen in the asteroid belt and on the comet, with its long blue solar wings. Then we send nanoprobes down through the metal clouds. In some regions the carnage was not as great and in one of these a probe detects a nearly whole satellite with a very rough surface, indicating that it was among the more ancient objects in this terrible sea (Vanguard 1). In the pictures the probe transmits, we see a dull grey sphere as pitted as the moon we'd just left, with six stubby legs. Its sphericity had saved it: there were no projections or joints to be knocked off by other debris, although it looked like the legs had once been longer antennas. In the little scratched panels on its surface we saw the precursors of the panes on the larger bird we'd captured. Our laser sensors pick up a reflection bouncing back from another spherical craft, spinning with diamond eyes (LAGEOS 1). Some reflectors are cracked and splice the laser beam into multiple directions, but a few return a clear signal to us.

Pop! A little sizzle indicates one of the lower nanoprobes has been pulverised, adding its dust to the miasma. But others have made it even lower, and are sending back pictures of a blue and white atmospheric surface. They also pick up radio signals; so we know there is someone or something down there. The images show a water world with several blocky continents. Geometric patterns of lights, structures and green vegetation are the archaeological signatures we'd been seeking; but we can't see much closer as the nanoprobes start to burn up. No matter; with the amount of data gathered, there is much we can infer.

It's hard to tear ourselves away, fascinated by the richness and tragedy of this planet. But there were two more planets to go. The next is obscured by clouds of the atmospheric type (Venus). We find nothing in orbit around it, and despite hopes that below the clouds were conditions suitable for life, it turns out to have nothing but a few landing craft. Some were crushed and others are standing upright, casting diffuse orange shadows over a stony land. They look so radically different from the rest of the solar system that we hypothesise manufacture by a distinct cultural group from the entombed planet. This place had been abandoned much longer than the blue planet's moon.

Closest to the sun, on another atmosphereless, heavily cratered body (Mercury), we see four craters which, judging by their sharp edges, were formed so recently in geological terms that they must have been the crash sites of orbiters. This was all; there was nothing else.

Our path back involves a swing around the sun. On the way, we intercept a heavily shielded spacecraft which could only have been sent out to measure the star (the Parker Solar Probe). It is silent too. Then we accelerate and leave this system behind us. We have much food for thought; it's clear that the spacefaring species on the blue planet was on the wrong side of an equilibrium between solar system use and orbital self-destruction.

This is the reverse of the journey I started with, both chronologically and geographically. Outside Earth, we may finally see what the characteristics of human nature really are. Which ones persist will be those we can't escape by reason of our biology and evolutionary history. They will be the basis of human evolution as a multigravity species. But I think we have a choice about which are supported by social and political structures, just as we do in the present world. Perhaps I could call the suite of behaviours which emerges from the human engagement with space 'behavioural futurity'. Different futures can't be imagined without understanding the diversity of the past. There are so many ways of connecting to space, and the stories we tell about space make a difference.

THE SMALL DANCE

Random meetings often bring the right kind of knowledge at the very time you need it. A conversation with a dancer at a space conference led me to something I'd never heard of before: the Small Dance. The Small Dance was devised by American choreographer Steve Paxton in the 1970s. It was built around gravity and balance, using the floor as another body to move against, a participant in the dance in its own right. It's the lowest possible orbit, where you are zero metres above the surface of Earth.

This is a simple version of the Small Dance, if you want to try it. Stand up, barefoot, and close your eyes. Make a very small nod of your head as if for *yes*, and very small shake as if for *no*. Relax your body and breathe into it, feeling your spine holding you up. Make the smallest stretch you can and rock your weight from foot to foot slightly. Feel everything inside your body, all the organs, and how gravity affects them. Be conscious of standing up and all the minute muscle adjustments your body makes to keep you upright. As Steve

Paxton said, 'You've been swimming in gravity since the day you were born. Every cell knows where down is. Easily forgotten. Your mass and the earth's mass calling to each other …'

When I stand and perform the Small Dance, I can feel the shifting of my body within itself like the faint movements of light falling through leaves. Time is layered beneath my feet in earlier land and sea surfaces, with the bones of extinct animals and humans who walked when the constellations were not those we know. Grains of cosmic dust are buried among the artefacts of long-forgotten human minds, all drawn downwards into the earth.

When you're ready let your body slowly relax into a squat, and just as slowly rise up again. This is the Small Dance.

But don't stop here. Look up. Imagine the space junk hundreds of kilometres above you, and the Moon with the silent shadows of the abandoned landing craft. Far beyond it, imagine Voyager silently singing with the voice of Waliparu, a human songline extending beyond the last planet, and the dust that circles past the spacecraft, as it drifts back towards Earth and falls into gravity, reminding us that the infinite is ever present, ready to embrace us in a tiny caress that leaves the smoothness of the cosmic between the fingertips and, perhaps, a sense of joy, that all of this is within reach, without ever leaving Earth.

SELECTED REFERENCES

INTRODUCTION

Clark, A, 2003, *Natural-born Cyborgs: Minds, Technologies, and the Future of Human Intelligence*, Oxford University Press, Oxford.

d'Errico, F and CB Stringer, 2011, 'Evolution, revolution or saltation scenario for the emergence of modern cultures?', *Philosophical Transactions of the Royal Society B: Biological Sciences*, 366, pp. 1060–1069.

Gorman, AC, 2011, 'The sky is falling: How Skylab became an Australian icon', *Journal of Australian Studies*, 35(4), pp. 529–546.

LeGuin, UK, 1973, *The Left Hand of Darkness*, Panther, St Albans.

Pop, V, 2011, 'Space exploration and folk beliefs on climate change', *Astropolitics*, 9(1), pp. 50–62.

Rifkin, RF, CS Henshilwood and M Mathison Haaland, 2015, 'Pleistocene figurative art *mobilier* from Apollo 11 cave, Karas Region, Southern Namibia', *South African Archaeological Bulletin*, 70(201), pp. 113–123.

Verne, J, 1865, *De la Terre à la Lune: Trajet Direct en 97 Heures*, Pierre-Jules Hetzel, Paris.

Wynn, T and FL Coolidge, 2011, 'The implications of the working memory model for the evolution of modern cognition', *International Journal of Evolutionary Biology*, pp. 1–12.

CHAPTER 1

Ackerman, LM, 1958, 'Facetious variations of "Sputnik"', *American Speech*, 33(2), pp. 154–156.

Cayley, NW and T Lindsey, 1984, *What Bird is That?* (revised by Terence R Lindsey), Angus & Robertson, Sydney.

Clarkson, C, Z Jacobs, B Marwick, R Fullagar, L Wallis, M Smith, RG Roberts, E Hayes, K Lowe, X Carah, SA Florin, J McNeil, D Cox, LJ Arnold, Q Hua, J Huntley, HEA Brand, T Manne, A Fairbairn, J Shulmeister, L Lyle, M Salinas, M Page, K Connell, G Park, K Norman, T Murphy and C Pardoe, 2017, 'Human occupation of northern Australia by 65,000 years ago', *Nature*, 547(7663), pp. 306–310'.

Cooper, HM, 1948, *Australian Aboriginal Words and Their Meanings*, South Australian Museum, Adelaide.

Selected references

Deetz, James, 1967, *Invitation to Archaeology*, Natural History Press, New York.

Ellinghaus, K, 1997, 'Racism in the never-never: Disparate readings of Jeannie Gunn', *Hecate*, 23(2), pp. 76–94.

Fewer, G, 2002, 'Towards an LSMR and MSMR (Lunar and Martian Sites and Monuments Records): Recording the planetary spacecraft landing sites as archaeological monuments of the future' in M Russell (ed.), *Digging Holes in Popular Culture: Archaeology and Science Fiction*, pp. 112–172, Oxbow Books, Oxford.

Garnaut, C, R Freestone and I Iwanicki, 2011, 'Cold War heritage and the planned community: Woomera Village in outback Australia', *International Journal of Heritage Studies*, 18(6), pp. 1–23.

Gill, ED, 1966, 'Provenance and age of the Keilor cranium: Oldest known human skeletal remains in Australia', *Current Anthropology*, 7(5), pp. 581–584.

Gorman, AC, 2000, 'The Archaeology of Body Modification: The Identification of Symbolic Behaviour Through Usewear and Residues on Flaked Stone Tools', PhD thesis, Department of Archaeology and Palaeoanthropology, University of New England, Armidale.

Gorman, AC, 2009, 'The archaeology of space exploration' in D Bell and M Parker (eds), *Space Travel and Culture: From Apollo to Space Tourism*, Blackwell Publishing, Malden, pp. 132–145.

Gunn, Mrs A, 1962, *Little Black Princess of the Never-Never*, Angus & Robertson, Sydney.

Lockwood, D, 1963, *We, the Aborigines*, Cassell Australia, Melbourne.

O'Leary, BL, 2009, 'One giant leap: Preserving cultural resources on the moon' in A Darrin and B O'Leary (eds), *The Handbook of Space Engineering, Archaeology and Heritage*, pp. 757–780, CRC Press, Boca Raton.

Rathje, W, 1999, 'An archaeology of space garbage', *Discovering Archaeology*, October, pp. 108–122.

Rathje, W and C Murphy, 2001, *Rubbish! The Archaeology of Garbage*, new edn, University of Arizona Press, Tucson.

Stewart, Susan, 1981, *Yellow Stars and Ice*, Princeton University Press, Princeton.

The Burra Charter: The Australia ICOMOS Charter for Places of Cultural Significance, 2013.

Van Loon, HW, 1922, *Ancient Man: The Beginnings of Civilizations*, Boni and Liveright Inc., New York.

CHAPTER 2

Asimov, Isaac, 1954, *Lucky Starr and the Oceans of Venus*, Doubleday and Co., New York.

Bédard, DG, A Wade, D Monin and R Scott, 2010, 'Spectrometric Characterization of Geostationary Satellites', paper presented to the AMOS (Advanced Maui Optical and Space Surveillance Technologies) conference, Hawaii, 2010. Available online: <www.amostech.com/TechnicalPapers/2012/POSTER/BEDARD.pdf>.

Freud, S, 1919, 'The Uncanny (Das Unheimliche)', *Imago* V: S 297–324.

Green, CM and M Lomask, 1970, *Vanguard: A History*, The NASA Historical Series, NASA SP-4202, National Aeronautics and Space Administration, Washington, DC.

Hammer, P and O Mace, 1970, 'Student satellite in orbit', *University of Melbourne Gazette*, 26(2), pp. 1–3.

Hays, PL and CD Lutes, 2007, 'Towards a theory of spacepower', *Space Policy*, 23, pp. 206–209.

Jones, DW, 1970, 'Oscar Five', *Newtrino*, 3(2), pp. 44–45.

Lewis, CS, 1943, *Voyage to Venus (Perelandra)*, The Bodley Head, London.

McCray, PW, 2008, *Keep Watching the Skies! The Story of Operation Moonwatch and the Dawn of the Space Age*, Princeton University Press, Princeton.

Mace, O, 2017, *Australis Oscar 5: The Story of How Melbourne University Students Built Australia's First Satellite*, ATF Press, Adelaide.

Morton, P, 1989, *Fire Across the Desert: Woomera and the Anglo-Australian Joint Programme, 1946–80*, Australian Government Publishing Service, Canberra.

Musk, E, 2017, Twitter account, 3 December, <twitter.com/elonmusk/status/937041986304983040>.

O'Brien, B, 2018, 'Paradigm shifts about dust on the Moon: From Apollo 11 to Chang'e-4', *Planetary and Space Science*, 156, pp. 47–56.

Pynchon, T, 1973, *Gravity's Rainbow*, Viking Press, New York.

Rueter, C, 2015, 'Our trajectory', TychoGirl blog, <tychogirl.wordpress.com/2015/04/22/our-trajectory/>.

Turner, V, 1967, *The Forest of Symbols: Aspects of Ndembu Ritual*, Cornell University Press, Ithaca.

Zubritsky, E, 2016, 'Now 40, NASA's LAGEOS Set the Bar for Studies of Earth', NASA History website, <www.nasa.gov/feature/goddard/2016/now-40-nasas-lageos-set-the-bar-for-studies-of-earth>.

CHAPTER 3

Burrows, WE, 1999 [1998], *This New Ocean: The Story of the First Space Age*, Modern Library, New York.

Dickson, P, 2001, *Sputnik: The Shock of the Century*, Berkley Books, New York.

Dougherty, K, 2017, *Australia in Space*, ATF Press, Adelaide.

Edgeworth, M, 2010, 'Beyond human proportions: Archaeology of the mega and the nano', *Archaeologies*, 6(1), pp. 138–149.

Gorman, AC, 2018, 'Gravity's playground: Dreams of spaceflight and the rocket park in Australian culture' in D Jordan and R Bosco (eds), *Defining the Fringe of Contemporary Australian Archaeology: Pyramidiots, Paranoia and the Paranormal*, pp. 92–107, Cambridge Scholars Publishing, Newcastle-upon-Tyne.

Gorman, AC, 2007, 'La terre et l'espace: Rockets, prisons, protests and heritage in Australia and French Guiana', *Archaeologies*, 3(2), pp. 153–168.

Gorman, AC, 2016, 'Tracking cable ties: Contemporary archaeology at a NASA satellite tracking station' in UK Frederick and A Clarke (ed.), *That Was Then, This Is Now: Contemporary Archaeology and Material Cultures in Australia*, pp. 101–117, Cambridge Scholars Publishing, Newcastle-upon-Tyne.

Ings, S, 1998, 'Open Veins' in G Dozois (ed.), *The Mammoth Book of New SF 11*, pp. 544–558, Robinson, London.

Pierce, C, 2017, 'Planet', *The Southern Review*, 53(2), pp. 335–336.

Popular Mechanics, 1959, 'Playground rocket ship has three-story cages', *Popular Mechanics*, July, p. 133.

Southall, I, 1962, *Woomera*, Angus & Robertson, Sydney.

Selected references

CHAPTER 4

Adams, D, 2005, *The Hitchhiker's Guide to the Galaxy*, Gollancz, London.

Bergin, Thomas G, 1971, 'For a space prober', in HE Landsberg and J Van Mieghen (eds), *Advances in Geophysics*, vol 15, p. 136, Academic Press, New York and London.

Bondeson, J, 2001, *Buried Alive: The Terrifying History of Our Most Primal Fear*, WW Norton, New York.

Dembling, PG and SS Kalsi, 1973, 'Pollution of man's last frontier: Adequacy of present space environment law in preserving the resource of outer space', *Netherlands International Law Review*, 20, pp. 125–146.

Egginton, W, 1999, 'On Dante, hyperspheres and the curvature of the Medieval cosmos', *Journal of the History of Ideas*, 60(2), pp. 195–216.

Gorman, AC, 2005, 'The Archaeology of Orbital Space' in Proceedings of the 5th Australian Space Science Conference 2005, Melbourne, VIC, RMIT University, Melbourne, pp. 338–357.

Gorman, AC, 2009, 'The gravity of archaeology', *Archaeologies*, 5(2), pp. 344–359.

Kessler, DJ and BG Cour-Palais, 1978, 'Collision frequency of artificial satellites: The creation of a debris belt', *Journal of Geophysical Research*, 83(A6), pp. 2637–2646.

Koyré, A, 1957, *From the Closed World to the Infinite Universe*, Johns Hopkins Press, Baltimore.

Lewis, CS, 1938 [1968], *Out of the Silent Planet*, Pan Books, London.

NASA Johnson Space Center Oral History Project 2004, BG Cour-Palais interviewed by J Ross-Nazzal, Canyon Lake, Texas, 1 March 2004, NASA website: <www.jsc.nasa.gov/history/oral_histories/Cour-PalaisBG/Cour-PalaisBG_3-1-04.htm>.

Power, E and A Keeling, 2018, 'Cleaning up Cosmos: Satellite Debris, Radioactive Risk, and the Politics of Knowledge in Operation Morning Light', *The Northern Review*, 48, pp. 81–109.

Rand, LR, 2016, 'Orbital Decay: Space Junk and the Environmental History of Earth's Planetary Borderlands', Unpublished PhD dissertation, University of Pennsylvania, Department of History and Sociology of Science.

Sable, ML, 2010, 'Space Junk', Strange Horizons website: <strangehorizons.com/poetry/space-junk/>.

Schroeder, K, 2010, 'Laika's Ghost' in Jonathan Strahan (ed.), *Engineering Infinity*, Rebellion Publishing, Oxford.

CHAPTER 5

Capelotti, PJ, 2010, *The Human Archaeology of Space: Lunar, Planetary and Interstellar Relics of Exploration*, McFarland Publishers, Jefferson.

de Bergerac, C, 1657 [1899], *A Voyage to the Moon*, translated by A Lovell, Doubleday and McCLure Co., New York, Gutenberg eBook: <www.gutenberg.org/files/46547/46547-h/46547-h.htm>.

Gaul, AT, 1956, *The Complete Book of Space Travel*, OH World Publishing Company, Cleveland.

Gaul, AT, Smithsonian Folkways Recordings 2018, 'Sounds of Insects': <folkways.si.edu/sounds-of-insects/album/smithsonian>.

Godwin, F, 1638, *The Man in the Moone or A Discourse of a Voyage Thither by Domingo*

Gonsales, printed by John Norton, for Joshua Kuton and Thomas Warren, London.

Gorman, AC, 2014, 'The Anthropocene in the Solar System', *Journal of Contemporary Archaeology*, 1(1), pp. 89–93.

Gorman, AC and BL O'Leary, 2007, 'An ideological vacuum: The Cold War in outer space' in J Schofield and W Cocroft (eds), *A Fearsome Heritage: Diverse Legacies of the Cold War*, Left Coast Press, Walnut Creek, California, pp. 73–92.

Harrison, TP, 1954, 'Birds in the Moon', *Isis*, 45(4), pp. 323–330.

NASA 2011, 'NASA's Recommendations to Space-Faring Entities: How to Protect and Preserve the Historic and Scientific Value of US Government Artefacts', Human Exploration and Operations Mission Directorate, Strategic Analysis and Integration Division, NASA.

O'Leary, BL, 2015, '"To boldly go where no man [sic] has gone before": Approaches in space archaeology and heritage' in BL O'Leary and PJ Capelotti (eds), *Archaeology and Heritage of the Human Movement into Space*, pp. 1–12, Springer, Heidelberg.

Rueter, Christine, 2015, 'Color Leaked In', TychoGirl blog, <tychogirl.wordpress. com/2015/04/19/color-leaked-in/>.

Slaughter, JW, 1902, 'The Moon in Childhood and Folklore', *The American Journal of Psychology*, 13(2). pp. 294–318.

Snow, CP, 1959, *The Two Cultures*, the Rede Lecture, Cambridge University Press, Cambridge.

Tanizaki, J, 1977 [1933], *In Praise of Shadows*, translated by TJ Harper and EG Seidensticker, Leete's Island Books, Stony Creek.

West, RH, 1895, 'Flight of birds across the Moon's disc', *Nature*, 53(1363), p. 131.

Westwood, L, BL O'Leary and MW Donaldson, 2017, *The Final Mission: Preserving NASA's Apollo Sites*, University Press of Florida, Gainesville.

Wolfgang von Goethe, J, 1840 [1810], *Goethe's Theory of Colours*, translated by CL Eastlake, John Murray, London.

CHAPTER 6

Berger, John, 1973, *Ways of Seeing*, London, BBC and Penguin Books.

Doolittle, H (HD), 1982, 'Eurydice', *Collected Poems 1912–1944*, New Directions Publishing Corporation, New York.

Harmand, S, JE Lewis, CS Feibel, CJ Lepre, S Prat, A Lenoble, X Boës, RL Quinn, M Brenet, A Arroyo, N Taylor, S Clément, G Daver, J-P Brugal, L Leakey, RA Mortlock, JD Wright, S Lokorodi, C Kirwa, DV Kent and H Roche, 2015, '3.3-million-year-old stone tools from Lomekwi 3, West Turkana, Kenya', *Nature* 521, pp. 310–315.

Jenkins, D and J Schofield, 2015, 'A journey to the heart of matter: The physical and metaphysical landscapes of CERN', *Landscapes*, 16(1), pp. 79–96.

Le Brun Holmes, S, 1999, *Faces in the Sun: Outback Journeys*, Viking, Ringwood.

Morena, AM, 2016, 'How a NASA Record Is Exploiting Indigenous Performers', the *Establishment*, 21 October: <medium.com/the-establishment/how-a-nasa-record-is-exploiting-indigenous-performers-5e806c09dad1>.

Sagan, C, 1978, *Murmurs of Earth: The Voyager Interstellar Record*, Random House, New York.

CHAPTER 7

Bradbury, R, AC Clarke, B Murray, C Sagan and W Sullivan, 1973, *Mars and the Mind of Man*, Harper & Row Publishers, New York.

Carter, P, 1987, *The Road to Botany Bay: An Exploration of Landscape and History*, University of Minnesota Press, Minneapolis.

Freud, S, 1989 [1913], *Totem and Taboo*, WW Norton, London.

Gorman, AC, 2009, 'Beyond the Space Race: The material culture of space in a new global context' in C Holtorf and A Piccini (eds), *Contemporary Archaeologies: Excavating Now*, Peter Lang, Frankfurt, pp. 161–180.

Gorman, AC, 2005, 'The cultural landscape of interplanetary space', *Journal of Social Archaeology*, 5(1), pp. 85–107.

Lewis, CS, 1944, *Voyage to Venus (Perelandra)*, The Bodley Head, London.

Loewenstein, G, 1994, 'A psychology of curiosity: A review and reinterpretation', *Psychological Bulletin*, 116(1), pp. 75–98.

Messeri, L, 2016, *Placing Outer Space: An Earthly Ethnography of Other Worlds*, Duke University Press, Durham.

Musk, E, 2018, Twitter account, 24 June: <twitter.com/elonmusk/status/1011083630301536256>.

Musk, E, Twitter account, 2018, 24 June: <twitter.com/elonmusk/status/1011085383239589888>.

Parker, KL, 1896, *Australian Legendary Tales*, David Nutt, London and Melvin, Mullen and Slade, Melbourne.

Platoff, AM, 1993, *Where No Flag Has Gone Before: Political and Technical Aspects of Placing a Flag on the Moon*, NASA Contractor Report 188251, prepared for Lyndon B. Johnson Space Center.

Plumwood, V, 1993, *Feminism and the Mastery of Nature*, Routledge, London.

Savage, MT, 1992, *The Millennial Project: Colonizing the Galaxy in Eight Easy Steps*, Little, Brown and Company, Boston.

Shōnagon, S, 1971, *The Pillow Book of Sei Shōnagon*, translated and edited by Ivan Morris, Penguin, New York.

Van der Krogt, P and F Ormeling, 2014, 'Michiel Florent van Langren and lunar naming', *Els noms en la vida quotidiana: Actes del XXIV Congrés Internacional d'ICOS sobre Ciències Onomàstiques,* Annex, Section 8, pp. 1851–1868.

CHAPTER 8

Ackerman, S, 2017, 'Exploring Freud's resistance to The Oceanic Feeling', *Journal of the American Psychoanalytic Association*, 65(1), pp. 9–31.

Lem, S, 1971, *The Futurological Congress*, Seabury Press, New York.

Melville, H, 1851, *Moby-Dick; or The Whale*, Harper and Brothers Publishers, New York and Richard Bentley, London.

Merati, G, S Rampichini, M Roselli, E Roveda and G Pizzini, 2006, 'Gravity and gravidity: Will microgravity assist pregnancy?', *Sports Sciences for Health*, 1(3), pp. 129–136.

Morton, T, 2010, *The Ecological Thought*, Harvard University Press, Cambridge MA.

Nin, Anaïs, 1972, speech at Hampshire College, Amherst, Massachusetts, available on YouTube: <www.youtube.com/watch?v=P592zVGFA0Q>.

Noordung, H, 1929, *The Problem of Space Travel: The Rocket Motor*, E Stuhlinger and
    JD Hunley (eds) with J Garland, NASA SP-4206, Washington DC.

Oman-Reagan, MP, 2015, 'Unfolding the space between stars: Anthropology of the
    interstellar', SocArXiv, Open Science Framework, manuscript, submitted
    4 February 2017: <osf.io/preprints/socarxiv/r4ghb/>.

Paxton, S, 1975, 'Contact improvisation', *The Drama Review*, 19(1), pp. 40–42.

Smith, C, 1975, *Norstrilia*, Ballantine Books, New York.

Smith, C, 1950, 'Scanners live in vain', *Fantasy Book*, 1(6), pp. 32–73, 85–88.

Turner, R, 2010, 'Steve Paxton's "Interior Techniques": Contact improvisation and
    political power', *TDR/The Drama Review*, 54(3), pp. 123–135.

INDEX

Index

Index